Advanced Techniques for Maintenance Modeling and Reliability Analysis of Repairable Systems

Scrivener Publishing
100 Cummings Center, Suite 541J
Beverly, MA 01915-6106

Publishers at Scrivener
Martin Scrivener (martin@scrivenerpublishing.com)
Phillip Carmical (pcarmical@scrivenerpublishing.com)

Advanced Techniques for Maintenance Modeling and Reliability Analysis of Repairable Systems

Garima Sharma
Senior Reliability Engineer, Valeo India Pvt. Ltd., Chennai, India
and
Rajiv Nandan Rai
Subir Chowdhury School of Quality and Reliability, Indian Institute of Technology, Kharagpur, India

Scrivener Publishing

WILEY

This edition first published 2023 by John Wiley & Sons, Inc., 111 River Street, Hoboken, NJ 07030, USA and Scrivener Publishing LLC, 100 Cummings Center, Suite 541J, Beverly, MA 01915, USA

For more information about Scrivener publications please visit www.scrivenerpublishing.com.

Wiley Global Headquarters
111 River Street, Hoboken, NJ 07030, USA

For details of our global editorial offices, customer services, and more information about Wiley products visit us at www.wiley.com.

Library of Congress Cataloging-in-Publication Data

ISBN 978-1-394-17443-0

Cover image: Pixabay.Com
Cover design by Russell Richardson

Set in size of 11pt and Minion Pro by Manila Typesetting Company, Makati, Philippines

Printed in the USA

10 9 8 7 6 5 4 3 2 1

Contents

List of Figures

List of Tables

Preface

The necessity of reliability and maintenance modeling and analysis of repairable systems has been showing an increasing trend due to the complexity of such systems. Repairable systems are the systems that can be restored to an operating condition after failure by any method other than the replacement of whole system. Moreover, they undergo various preventive maintenances (PM) and are often subjected to imperfect maintenance. To undertake the reliability analysis of repairable systems, most of the industries establish their own set up of maintenance, repair, and overhaul (MRO) facilities.

Before laying hands on this book, the authors realized that though various models and maintenance policies are available for the repairable systems, they still lack in providing the solutions to the real-time industrial problems. The models and maintenance policies that can be directly implemented are to be developed. The authors then undertook the task of developing such reliability models and maintenance policies that could be very much useful and beneficial for conducting advanced reliability analysis of repairable systems.

This book provides a framework for the modeling and analysis of repairable systems considering parametric estimation of the failure data. The book provides due exposure to the advanced generalized renewal process (GRP) based models and maintenance policies along with its applications to repairable systems data from aviation industry. The book also covers the dependency modeling of various factors that affect the repair effectiveness of the system. This text is intended to be useful for senior undergraduate, graduate and post graduate students in engineering schools and also for professional engineers, reliability administrators and managers.

This text has primarily emerged from the industrial and research experience of the authors. A number of illustrations have been included to make the subject pellucid and vivid especially to the readers who are new to this area. Various examples have been provided to showcase the applicability of presented models and methodologies to assist the reader in applying the concepts presented in this book.

The basic concepts of reliability analysis of repairable systems and generalized renewal process (GRP) can readily be seen in various available texts that deal with reliability analysis of repairable systems. The reliability literature is plentiful covering such aspects in reliability data analysis where the failure times are modeled by appropriate life distributions. Hence, the reader is advised to refer to any such textbook on repairable systems reliability analysis for a better comprehension of this book.

Chapter 1 introduces the reader to the basics of maintenance, repair and overhaul (MRO). by providing a preview of the fundamental understanding of MRO, so that it will be easier to integrate the knowledge of the advanced reliability techniques in dealing with reliability analysis of repairable systems, which forms the main theme of this book.

Chapter 2 provides an overview of basic concepts and the terminologies used in repairable system reliability analysis, types of maintenance along with the GRP based imperfect maintenance models available in the literature.

Chapter 3 presents a virtual age model considering scheduled preventive maintenance (SPM), corrective maintenance (CM) and overhaul as imperfect. The chapter also describes the activities done during SPM and overhaul along with the need for imperfect overhaul assumption. The model is demonstrated with the help of field failure data of aero engines.

Chapter 4 explains a technique and virtual age models to deal with failure modes (FM) wise censored data of repairable systems. The chapter also highlights the need for this particular technique in industries.

Chapter 5 presents a methodology for risk-based threshold on intensity function for the identification of high failure rate components (HFRCs). The chapter brings out the limitations of the present methods of HFRC designation.

Chapter 6 provides a maintenance policy called progressive maintenance policy (PMP) for large and critical repairable systems. The policy is demonstrated using the same data set as utilized in chapter 5. The superiority of the policy is discussed by comparing it with the conventional maintenance policies followed by the aviation industries in the Indian context.

Chapter 7 presents an age-based maintenance policy to obtain the optimal overhaul time for repairable systems considering CM and SPM and imperfect. The chapter also provides the spare parts estimation model based on virtual age concept.

Chapter 8 brings out the limitations of quantitative assessment over the qualitative assessment of repair effectiveness index (REI) and presents various subjective factors which can affect REI the most. As an example, a dependency model with the help of the Bayesian network (BN) is developed to model their inter-dependency with each other and REI.

The book makes an comprehensive attempt to provide a coverage to various models and methodologies that can be used for advanced modeling and analysis of repairable systems reliability analysis. However, there is always a scope for improvement and we are looking forward to receiving critical reviews and/or comments of the book from students, teachers, and practitioners. We hope readers will all gain as much knowledge, understanding and pleasure from reading this book as we have from writing it.

Garima Sharma
Rajiv Nandan Rai

1

Maintenance, Repair, and Overhaul: A Preview

1.1 Introduction

To undertake the reliability analysis of repairable systems, most of the industries establish their own setup of maintenance, repair, and overhaul (MRO) facilities. Before we embark upon the remaining contents of the book, the authors thought it imperative to introduce the readers to the basics of maintenance, repair, and overhaul. This chapter provides a preview of the fundamental understanding of MRO so that it will be easier to assimilate the comprehension of the advanced reliability techniques to deal with repairable systems, which has been endeavored in this book.

Maintenance, repair, and overhaul, or MRO, provides life cycle maintenance through routine preventive maintenance, planned out-of-service maintenance, or (corrective) repairs, overhaul, or rebuilds for damaged equipment. Even though industries account for the majority of them, a product or piece of equipment with high costs and a long lifespan is definitely a candidate for MRO services. Examples include massive manufacturing machinery, electric power generation, marine boats and infrastructure, mass transit vehicles, military vehicles, and systems.

Industrial systems generally deteriorate over time due to use and exposure to environmental factors. This deterioration eventually results in system failure, which in turn causes safety problems, equipment damage, quality problems, and unplanned machine downtime. A few decades ago, maintenance was mainly thought of as something challenging to manage and had to be done after a failure. Maintenance is widely acknowledged as a crucial component of asset management and a crucial commercial function. Organizations are becoming more aware of how maintenance intervention planning may increase their productivity and reliability. Preventive maintenance activities increase as a result and better fit with other business processes like production scheduling and spare parts management.

Garima Sharma and Rajiv Nandan Rai. Advanced Techniques for Maintenance Modeling and Reliability Analysis of Repairable Systems, (1–36) © 2023 Scrivener Publishing LLC

For instance, companies in the process and chemical sectors can significantly boost profitability by preventing unscheduled stoppages. The continuous automation of production processes and an intensifying level of competition in the market have increased awareness of the need for good maintenance planning.

Keeping facility equipment, tools, and infrastructure in good condition and operating them efficiently is the objective of anybody who works in maintenance. This helps to prevent unanticipated downtime or equipment failure. This is what repair and maintenance allow us to do. Although the terms repair and maintenance are sometimes used interchangeably, they have various meanings in the asset management industry. When an asset breaks, is damaged, or ceases to function, repairs are restorative work that must be done. Routine tasks and/or corrective or preventive repairs performed on assets to avoid damage and extend life expectancy are referred to as maintenance. Examples include routinely cleaning grease traps, and air conditioning units, painting, and inspections.

An overhaul is a general maintenance procedure carried out on a piece of machinery or other industrial equipment. The purpose of an overhaul is to maintain the system's function. Regular inspections can stop a variety of critical damage. Typically, maintenance service providers carry out machinery overhauls. It is possible to agree on the frequency of overhauling; regular maintenance is typically planned for once a year. A more frequent equipment check is advised for older machinery, especially larger ones with complicated mechanics. Typically, overhauls begin with a thorough inspection. The overhauled machine is examined by skilled maintenance personnel. It indicates that the machine's operation is tracked while it is in use. The item of equipment should be disassembled following the initial inspection. For additional inspection and the subsequent overhauling stages, such as repair, disassembly is essential. An efficient machine breakdown by a trained maintenance technician can reveal which equipment components require replacement or repair. The machine is either fixed or certain damaged parts are changed, depending on the problem. This procedure demonstrates once more how efficient overhauling is compared with complete equipment replacement.

Since the spare parts may need to be acquired from a manufacturer, part replacement may require more time than a straightforward repair. Reassembling the entire mechanism is done after the successful substitution of spare parts. The reassembly, which comes last, is essential to the equipment's operation. Reassembly requires a certain level of competence; therefore, it is best left to the experts. It is difficult to determine whether a repair was successful without testing. Testing determines if the reassembly is successful or not; otherwise, the procedure restarts at the beginning (inspection).

In the further section, a brief understanding of the maintenance, repair, and overhaul is provided for the benefit of the readers, and the same is further amplified in the next chapter as per the subject covered in subsequent chapters.

1.2 Maintenance

Maintenance is an essential integral of any organization. It is far beyond just fixing a failed system. It is an investment for the future and guarantees that the system will continue to perform reliably. It is defined as "Maintenance is a philosophy to utilize all physical sciences with a disciplined manner at all levels of the operation, which assures that critical capital asset will provide long-term and reliable performance." Taking such a narrow perspective, maintenance activities will be confined to the reactive tasks of repair actions or item replacement triggered by failures. Thus, this approach is known as reactive maintenance, breakdown maintenance, or corrective maintenance. "All operations aimed at keeping an object in, or restoring it to, the physical state considered necessary for the performance of its production function" is how Geraerds (1983) described maintenance in a more modern context. Of course, proactive actions like routine maintenance, periodic inspections, preventive replacements, and condition monitoring are also included in the scope of this expanded view.

Depending on how duties are distributed within the business, these maintenance activities may be split among different departments. For instance, in a company that uses total productive maintenance (TPM), the maintenance department handles significant repairs and overhauls, while the operating staff is in charge of routine service and periodic equipment inspection (Nakajima, 1988). The actual maintenance procedure may be exclusive to a particular facility, industry, or collection of issues. Identical product organizations may use different maintenance systems with varying levels of technological development and production size, and the various systems may function effectively. As a result, maintenance systems are created employing formal decision-making tools and processes along with experience and judgment. However, two important factors should be taken into account: a strategy that determines which level of the plant to perform maintenance on, thus outlining a structure that will support the maintenance, and second, planning that deals with daily decisions on what maintenance tasks to perform and providing the resources to carry out these tasks. One of the fundamental and essential elements of the maintenance is the maintenance management function (MMF). According to Figure 1.1 below, the MMF entails organizing, planning, controlling, and implementing maintenance activities.

There are significant factors that must be taken into account while designing the maintenance organization. The factors to consider are maintenance capacity, centralization versus decentralization, and insourcing versus outsourcing maintenance. The following aspects or factors significantly impact a maintenance organization's role within the plant or the entire organization:

- Type of business: such as whether it is labor-intensive, high-tech, producing goods, or providing services.
- Goals: These could include things like maximizing profits, expanding market share, and other industries' goals.
- Size and organizational structure.
- Organizational culture and the scope of the maintenance responsibilities assigned.

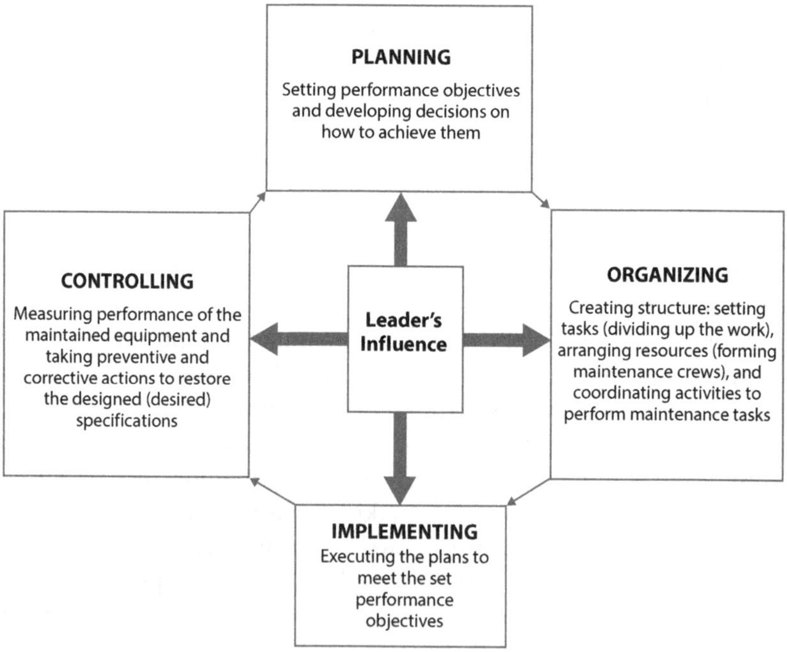

Figure 1.1 Activities in maintenance management function.

Maintenance expenses make up a sizeable amount of operating expenses for many asset-intensive enterprises. In larger businesses, a $1 million reduction in maintenance costs generates the same amount of profit as a $3 million increase in sales (Wireman T, 2007). Around 1,500 billion euros are spent annually on Europe's maintenance budget (Altmannshoffer R, 2006) and 20 billion euros per year for Sweden (Ahlmann H, 2002). In opencut mining, a typical dragline's loss of revenue ranges from US $0.5 to US $1 million per day. A 747 Boeing plane's loss of revenue ranges from US $0.5 to US $1 million per day (Murthy, Atrens, and Eccleston, 2002). As a result, business management is increasingly aware of the value of maintenance productivity.

Here are a few instances where inadequate maintenance procedures led to catastrophes and accidents that resulted in significant losses, including Bhopal, Piper Alpha, the Columbia space shuttle, and power outages in 2003 in New York, the UK, and Italy. Instead of saving a billion US dollars, such an accident may have been avoided and improved the organization's reputation. Maintenance performance monitoring has become a crucial component of strategic thinking for the manufacturing and service sectors. In order to oversee and monitor the maintenance process' performance, necessary and corrective steps must be taken to reduce and mitigate safety-related risks, uphold societal obligations, and improve the effectiveness and efficiency of the asset being maintained.

Maintenance performance is a metric frequently used by enterprises to gauge maintenance productivity. The ratio of a production system's output (the products or services delivered) to its input (the labor, materials, tools, plant, and equipment, among other things used to produce the products or services) is generally understood to be the definition of productivity. Reducing the amount of maintenance materials used as well as projects, outages, and overhaul costs would increase maintenance productivity (Wireman T, 2007). Optimizing maintenance resource use and boosting maintenance productivity are two ways to lower operation and production expenses (Duffuaa and Al-Sultan, 1997). In order to evaluate, monitor, control, and make suitable and timely decisions, evaluating maintenance productivity performance is essential for any production and operating firm.

What to measure, how to communicate maintenance performance throughout the business, and how to tie maintenance performance to goals and strategies are some of the issues that need to be clarified while creating maintenance metrics (Murthy, Atrens, and Eccleston, 2002). To achieve this, it is necessary to aggregate measured maintenance performance indicators, such as availability, reliability, and mean time between failures, from the shop floor level to the strategic levels for making management decisions. This basically entails breaking down the corporate objectives into measurable targets up to the shop floor level (A. H. c. Tsang, 2002).

The performance measurement needs to be viewed along three dimensions (Andersen B and Fagerhaug T, 2007), i.e.,

- Effectiveness,
- Satisfaction of customer needs,
- Efficient, economic, and optimal use of enterprise resources,
- Changeability—strategic awareness to handle changes.

The effectiveness of the plant's operations is significantly influenced by its maintenance practices and safety record. To satisfy supply deadlines, costs, quality standards, and quantity requirements, management must rely on the anticipated plant capacity. To reach ideal production levels, a suitable maintenance and safety approach must be modified. Some of the essential measures of maintenance productivity are the total cost of maintenance/total production cost, availability, production rate, quality rate, mean time to repair (MTTR), mean time between failures (MTBF), maintenance breakdown severity, maintenance improvement, maintenance cost per hour, manpower utilization, manpower efficiency, material usage/work order, and maintenance cost index.

Maintenance performance indicator (MPIs) are used to assess maintenance effectiveness (Wireman T, 1998). A product of numerous measurements is an indicator (measures). A performance indicator is a measurement that can produce a quantitative number to show the degree of performance while accounting for one or more aspects. In addition to many other uses, MPIs could be used for financial reporting; employee performance reviews; customer satisfaction surveys; health, safety, and environmental (HSE) ratings; and overall equipment effectiveness (OEE). In reliability maintenance studies, statistics and probability are essential tools.

In maintenance engineering, failure, function, and functionability of the system are important terms to understand. Failure is an event whose occurrence results in either loss of ability to perform the required function or loss of ability to satisfy the specified requirements. In other words, it is the inability to fulfill any of the component functionability. Functionability is defined as the inherent ability to perform a required function with specified performance and attributes when it is utilized as specified. Function is related purely to the function performed, whereas functionability takes into consideration the level of performance achieved. During its operational life, a restorable system that fluctuates between states of failure and

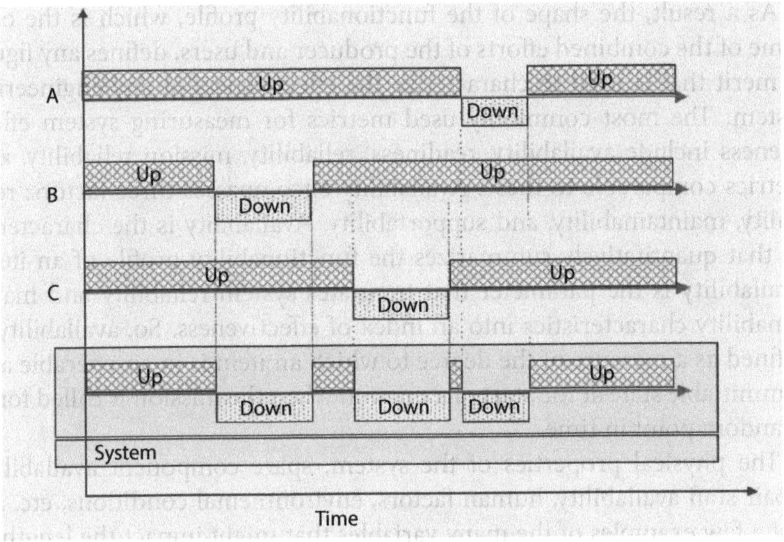

Figure 1.2 Example of a functionability profile in a series system.

function, an established pattern of mapping the states of the system during the utilization process, is known as the functionability profile as shown in Figure 1.2.

The pattern of a system's functionability profile, with a focus on the percentage of time the system in question will be available for functionability fulfillment, is thus one of the key concerns of users of any engineering system. It is obvious that the following two causes are primarily responsible for its particular shape:

1. Inherent characteristics: The decisions made by the designers and builders during the early stages of the system design determine a system's inherent qualities, such as reliability, maintainability, and supportability, which directly determine the frequency of failures, the complexity of maintenance tasks, and the ease of the support of the tasks required.
2. Utilization characteristics: Each user of the system determines its operational characteristics, which are driven by the operational scenario, maintenance policy, and logistics support concept, with the aim of managing the provision of the resources required for the effective completion of all operation and maintenance chores.

As a result, the shape of the functionability profile, which is the outcome of the combined efforts of the producer and users, defines any figure of merit that is used to characterize the effectiveness of any engineering system. The most commonly used metrics for measuring system effectiveness include availability, readiness, reliability, mission reliability, and metrics comparable to these. Availability encompasses three factors: reliability, maintainability, and supportability. Availability is the characteristic that quantitatively summarizes the functionability profile of an item. Availability is the parameter that translates system reliability and maintainability characteristics into an index of effectiveness. So, availability is defined as a measure of the degree to which an item is in an operable and committable state at the start of a mission when the mission is called for at a random point in time.

The physical properties of the system, spare component availability, repair staff availability, human factors, environmental conditions, etc. are just a few examples of the many variables that might impact the length of downtime. Based on the following criteria, downtime can be split into two categories:

- Waiting downtime: The device is currently inoperative but not being repaired at this time. This might be caused by the time it takes to process paperwork, ship replacement parts, etc.
- Active downtime: The equipment is currently undergoing repair during this time and is therefore not used, or to put it another way, active downtime is the amount of time it takes a repair or replacement to be made. Human variables, as well as the equipment's design, have a significant impact on how long the active downtime lasts.

Reliability is the probability/chance and capability of items (parts, components, equipment, subsystem, and system) to perform their intended function without failure for a desired period in specified use conditions (environment). It can be specified, allocated, predicted, designed-in, tested, estimated/predicted, and demonstrated with its maintainability, availability, safety, and quality level in an optimized way. On the other hand, maintainability is expressed as the probability that an item will be retained in or restored to a specific condition within a given period of time when the maintenance is performed in accordance with prescribed procedures and resources.

In maintenance, maintainability design is essential for any organization in steps as shown in Figure 1.3. It begins with defining maintainability

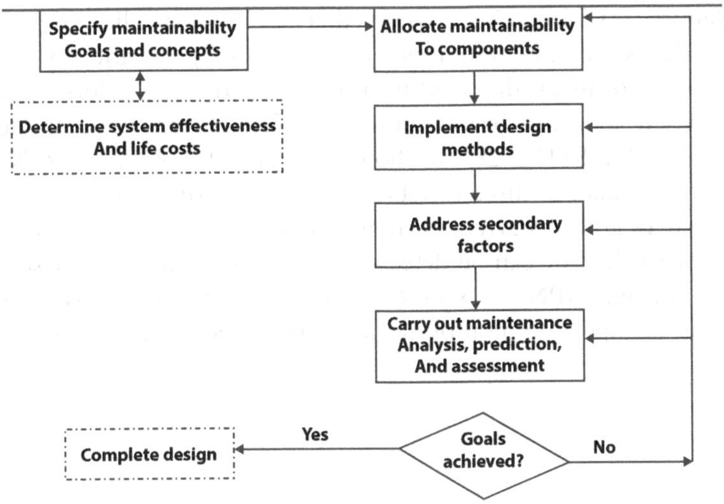

Figure 1.3 The maintainability design process.

goals. The determination of goals coincides with reliability specifications. The overall maintainability goals must be allocated to the repairable subassemblies. The maintainability design methods are applied to achieve the specified maintainability parameters. Secondary considerations that are included in this analysis are quantity and quality of resources. In the assessment of achieved maintainability, additional design activity may be necessary if goals are not attained.

There are several key concepts that should be followed as part of any design activity that supports this repair time reduction:

1. Fault isolation and self-diagnosis
2. Parts standardization and interchangeability
3. Modularization and accessibility
4. Repair versus replacement

The circumstances (that might not be disjoint) during the design, manufacturing, or use which led to the failure and useful information in the prevention of failures or their recurrence need to be recognized. Considering the consequences and liability of failures, it is important to establish reliability acceptance goals that will minimize and limit the occurrence of failures. Tools such as FMEA, FMECA, reliability growth, and reliability validation tests are important to ensure that the goals are integrated with the process and are tracked through the Design Verification Plan and Report (DVP&R). An industry that follows reliability engineering and standards utilizes failure mode and effect analysis/criticality analysis (FMEA/CA) for its product and/or process, where a risk priority number (RPN)-based approach has assumed precedence over the methodology described in MILSTD 1629A as shown in Figure 1.4. A classic RPN is an arithmetic product of three variables, namely, severity (how serious their consequences are), occurrences (how frequently they occur), and detection (how easily they can be detected in the design, process, or service), with the formula RPN = S × O × D. These three factors are constructed according to expert opinion or available data in a point scale of 10 or 5.

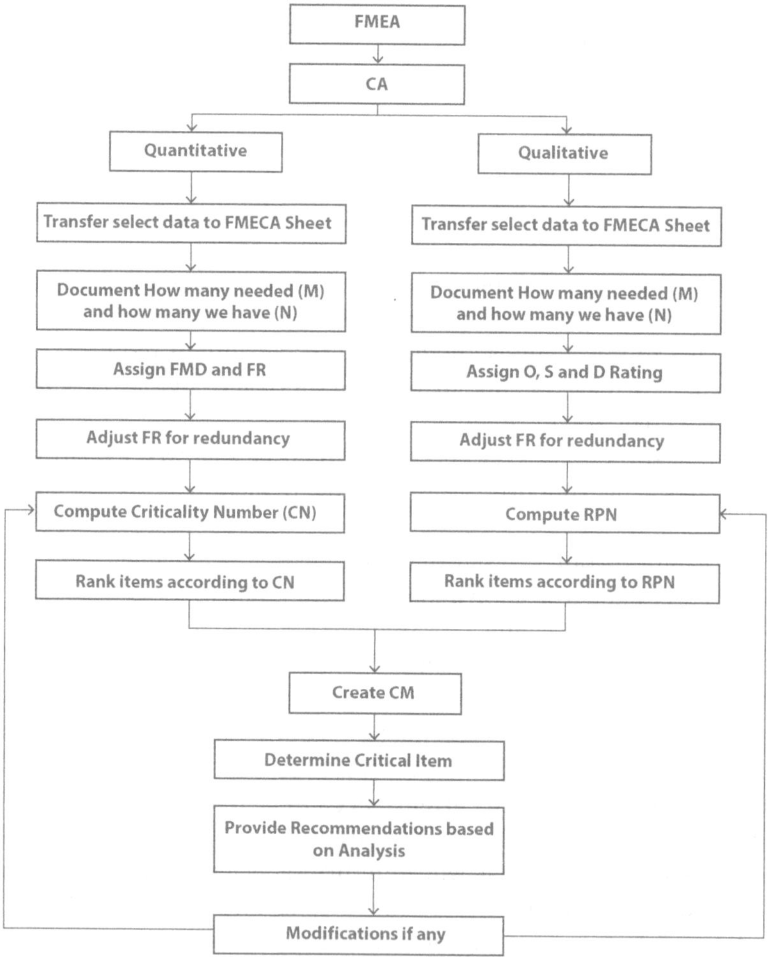

Figure 1.4 Flowchart of the FMEA/CA.

Failure modes refer to the potential failure types or ways. Failures can be possible or actual and include any error or flaws, especially those that negatively impact customers. Effects analysis is the study of the outcomes of those errors. Failures are ranked by the severity of their effects, frequency with which they occur, and ease with which they can be discovered. The goal of the FMEA is to eliminate or reduce failures by initiating action with the failures that have the highest priority, which can be determined by their RPN and/or factor(s). FMEA also records current understanding and actions on failure risks for use in ongoing improvement. FMEA is used in the design phase to prevent failures and, later, for control, both before and during the process operation. The standard FMEA technique has been expanded in a number of ways, but its core principles have not changed.

Reliability engineering is primarily concerned with fixing equipment issues that result in excessive production downtime and maintenance work. Engineering expertise is applied in reliability engineering for risk management, asset life cycle management, configuration management, asset maintenance, and loss prevention. The responsibilities of a maintenance engineer are more tactical, such as ensuring that the assets satisfy the company's current expectations. The maintenance engineer is in charge of day-to-day reliability duties, whereas a reliability engineer considers long-term reliability requirements. To increase the dependability of physical assets in the pursuit of continual improvement, two complementary approaches with various focuses are provided (uptime). These methodologies are as follows:

- Total productive maintenance (TPM)—a people-centered methodology.
- Reliability-centered maintenance (RCM)—an asset-centered methodology.

TPM, which reduces asset failure, production problems, and accidents, offers a thorough life cycle approach to asset management. Everyone in the company is involved, from senior management to production mechanics, production support teams, and external suppliers. The goal is to continuously increase an asset's availability and stop it from degrading so that it retains its optimum efficiency. Three categories, namely, autonomous maintenance, scheduled maintenance, and maintenance reduction, can be used to categorize TPM principles. The purpose of using asset operators autonomously is to carry out some standard maintenance tasks. These jobs involve the regular upkeep of the asset, such as cleaning, inspection, tightening, and lubricating. The maintenance personnel can begin working on proactive asset maintenance by automating some of the routine maintenance operations. The goal of planned maintenance activities, commonly referred to as preventive maintenance, is to replace worn-out parts and restore damaged assets. The maintenance cutbacks asset design and preventive maintenance, which aim to minimize the overall amount of maintenance necessary, make up the TPM idea. To reduce variation, boost productivity, cut maintenance costs, lower inventory, raise training, and improve safety are made possible through total productive maintenance. The suggestions for TPM enhancement include top-down guidance, process integration, and data-driven choices. It can be said that there are mainly eight pillars to achieve TPM as shown in Figure 1.5.

A *reliability-centered maintenance (RCM)* technique is used in reliability-based maintenance. The area covered by RCM includes defining system functions, defining functional and possible failures, exposing failure modes, recording failure impacts and consequences, identifying failure causes, and identifying tasks and activities to reduce the risk of failures and categories of tasks as described in Figure 1.6.

What kind of maintenance activities need to be conducted is determined by RCM, which also provides the answer to the question of what kind of maintenance strategies should be used on an asset. It is still unclear when to carry out the advised maintenance procedure in order to achieve

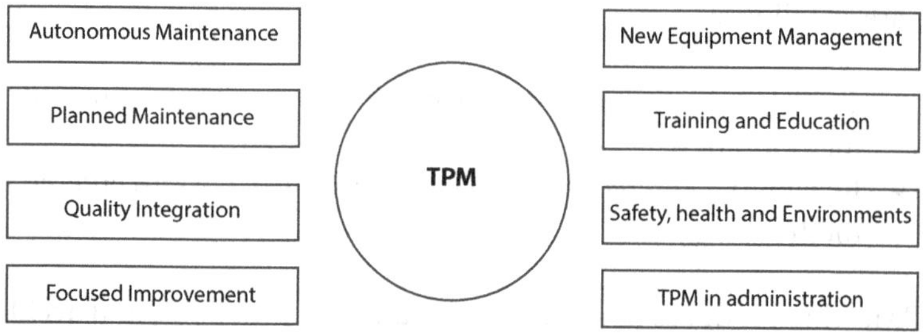

Figure 1.5 The pillars of TPM.

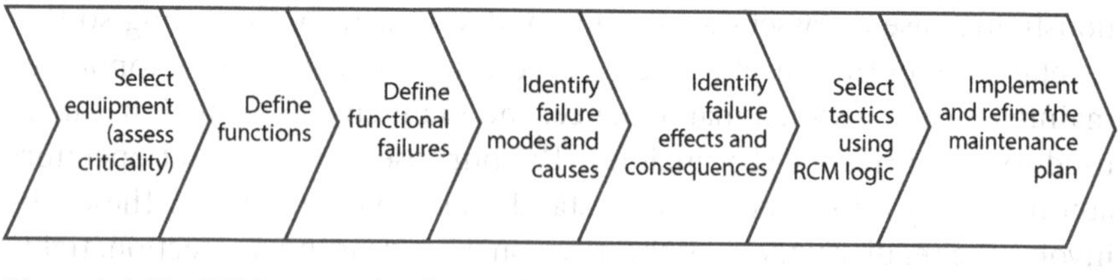

Figure 1.6 The RCM process (Jardine and Tsand, 2005).

optimal outcomes. Asset managers must take into account four crucial decision areas, which are shown as columns in Table 1.1, in order to maximize the life cycle value of the company's people and physical assets. Component replacement is the subject of the first column. Inspection

Table 1.1 Key areas of maintenance and replacement decisions (Jardine and Tsand, 2005).

Optimizing equipment maintenance and replacement decisions			
Component replacement	**Inspection procedures**	**Capital equipment replacement**	**Resource requirements**
1. Best preventive replacement time a) Deterministic performance deterioration b) Replace only on failure c) Constant interval d) Age-based 2. Spare parts provisioning 3. Repairable systems 4. Glasser's graphs 5. Software: SMS and OREST	1. Inspection frequency for a system a) Profit maximization b) Availability maximization 2. A, B, C, D class inspection intervals 3. FFIs for protective devices 4. Condition-based maintenance 5. Blended health monitoring and age replacement 6. Software: EXAKT	1. Economic life a) Constant annual utilization b) Varying annual utilization c) Technological improvement 2. Repair vs. replace 3. Software: PERDEC and AGE/CON	1. Workshop machines/ crew sizes 2. Right sizing equipment a) Own equipment b) Contracting out peaks in demand 3. Lease/buy 4. Software: workshop simulator and crew size optimizer
Probability and statistics (Weibull analysis including the software WeibullSoft)	Stochastic processes (for CBM optimization)	Time value of money (discounted cash flow)	Queuing theory simulation
Database (CMM/EAM/ERP system)			

activities, including condition monitoring, are the subject of the second column. Capital equipment replacement is the subject of the third column. The location of the resources needed for maintenance is the subject of the final column.

Replacement problems (and maintenance problems in general) can be classified as either deterministic or probabilistic (stochastic). Deterministic problems are those in which the timings and outcomes of the replacement actions are assumed to be known with certainty. For example, we may have an item that is not subject to failure but whose operating cost increases with use. Short-term and long-term deterministic models while considering minimum total cost and minimum downtime have been proposed by various authors, for example replacing the air filter in an automobile and overhauling a boiler plant. The timing and result of the replacement action depend on chance in probabilistic problems. In the simplest scenario, the equipment can be categorized as either working properly or not. The time distribution between the completion of the replacement activities and failures may be used to describe the probability law defining shifts from good to failure. Any series of events can be a replacement strategy. On the other hand, the best replacement policies are those that optimize or reduce a particular criterion, such as profit, total cost, and downtime, or guarantee that a specific safety or environmental criterion is not surpassed.

We now provide a glimpse of the concepts related to different types of maintenance for a better appreciation of the readers in this area. However, this book is mainly oriented toward providing advanced techniques on reliability analysis of imperfect maintenance related mainly with preventive and corrective maintenance.

Preventive, corrective, and predictive maintenances are the three typical categories of maintenance actions as shown in Figure 1.7. In order to keep equipment in working order, "preventive maintenance" refers to planned and routine tasks carried out on an operating system according to a set timetable. The term "corrective maintenance" describes procedures carried out as a result of equipment failure with the goal of getting the equipment back to a predetermined operational state. "Predictive maintenance" refers to maintenance tasks that are carried out based on "modern measurement and signal processing systems" condition diagnosis.

Several basic maintenance optimization models include the "age replacement policy" and the "block replacement policy"-based replacement models. These models take into account a technical system that is instantly replaced upon failure by a substitute system, incurring costs of $c > 0$ for each replacement and "penalty costs" of $k > 0$ for each system failure. This system has a lifetime described by a positive random variable T and

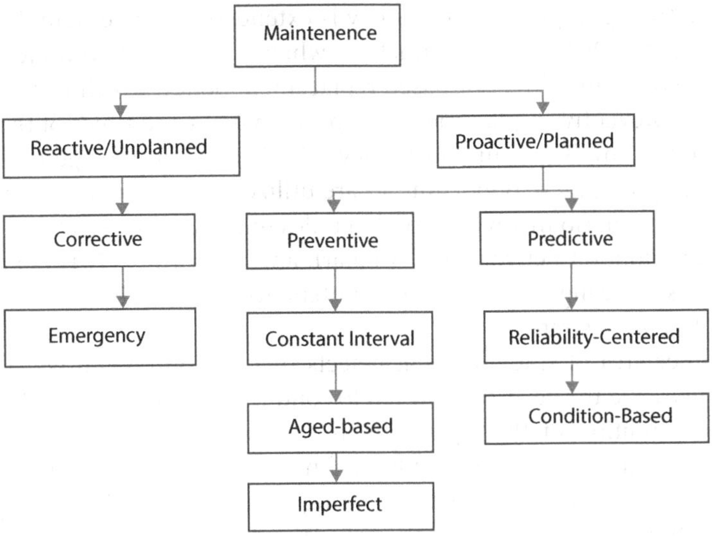

Figure 1.7 Types of maintenance.

a distribution *F*. The "age replacement policy" takes into account a fixed system replacement age that is indicated by a positive constant and aims to reduce the cost per unit of time throughout a series of system lifetimes. The average cost after *n* cycles is derived by dividing the total cost of every additional cycle by the number of additional system lifetimes. By dividing the overall cost per unit time that has occurred until time *t*, the model also takes consecutive cycle costs into account (or the current time). The system is replaced in accordance with this policy when it fails or when it reaches the specified system replacement age, whichever comes first. The "block replacement policy," in contrast, calls for preventive replacements to be carried out at predetermined intervals without taking into account the age of the system's components and at a cost of c. At the expense of $c + k$, the system is also replaced when it fails within the predetermined time frames. The block replacement policy, according to the authors, is simpler to manage because system/component replacements take place at predetermined times. The "age replacement policy," on the other hand, is regarded as being more pliable because it considers the system's age.

Jiang (2001) outlines a method for selecting an approximately ideal age-based maintenance strategy for a certain time frame. The ideal replacement age for a unit with a random and undetermined initial age is taken into account by Khatab *et al.* (2017). Dekker and Plasmeijer (2001) take into account maintenance opportunities that appear as a result of a Poisson process. The age replacement strategy is extended by Finkelstein, Shafiee, and Kotchap (2016) to the situation when the system's production is declining over time. The age-based replacement policy, which restricts the timing of preventive replacements to specified dates and times of the year, is introduced by Bajestani and Banjevic (2016). When the age reaches a particular point, these opportunities are utilized. Their model is used in a real-world scenario involving wooden poles in a Canadian electrical company's distribution network. He, Maillart, and Prokopyev (2017) examine the effects of negligent preventive maintenance and take into account the age replacement strategy.

Another area of study looks at defects that can only be discovered by *inspections*, and the key issue is when to conduct these inspections. Finding out the condition of the equipment is the main goal of an inspection. Depending on the state found, additional maintenance steps may need to be conducted once indicators, such as bearing wear, gauge readings, and product quality, which are used to characterize the state, have been specified and inspection was made to identify the values of these indicators. The expenses of the inspection (which will be tied to the indicators used to indicate the condition of the equipment) and the advantages of the examination, such as the discovery and adjustment of minor problems before a severe breakdown occurs, should dictate when the inspection should take place.

Three categories of inspection-related issues are typically looked at:

- Inspection frequencies: for continuously operating machinery that is prone to failures.
- Inspection intervals: for machinery only employed in emergency situations (failure finding intervals).
- Condition monitoring of equipment: optimizing condition-based maintenance (CBM) decisions.

Equipment maintenance and repairs can be expensive, time-consuming, and resource-intensive. Additionally, while the equipment is being fixed, industrial output declines. To reduce the frequency of breakdowns, we might frequently inspect the equipment and correct any minor issues that could later result in a complete breakdown. In terms of materials, labor expenses, and productivity loss due to scheduled downtime, these inspections are expensive. We must develop an inspection program that will provide us with the optimal balance between the frequency of inspections and the production in order to maximize the profit per unit time from the equipment over an extended period of time. Such a mechanism is seen in Figure 1.8. It may stop working for a number of reasons, including those

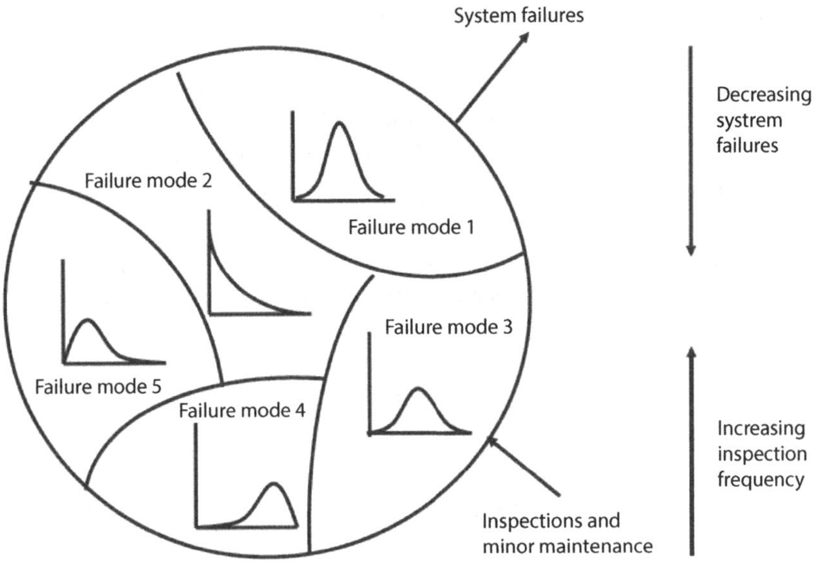

Figure 1.8 System failure (Jardine and Tsand, 2005).

caused by component 1, component 2, and so on. Each of these equipment failure reasons could have a unique independent failure distribution. It is believed that when inspection frequency or intensity increases, the frequency of equipment/system failures will decrease. It can be challenging to select the optimal frequency and intensity. Two models were suggested for this issue: one with the lowest cost and the other with the least amount of downtime.

Establishing an appropriate quantity of *emergency spares* that can be put into operation if a current long-life and highly dependable component breaks is a crucial aspect of spares management. Transformers in an electrical utility and electric motors in a conveyor system are two examples of such components. A few spare units might be maintained on hand to maintain a highly dependable service. The quantity of spares that should be stocked is an issue that needs to be answered in this section. It is necessary to indicate whether the spare part is one that is discarded after failure (a nonrepairable spare) or if it may be renewed and repaired after failure before being restocked in order to properly respond to the question (a repairable spare). Finally, it is important to comprehend the end result. For the purpose of determining the ideal number of both nonrepairable and repairable spares, four parameters will be taken into account in this section, which are as follows:

- Instantaneous reliability: Based on current circumstances, there is a probability that a spare is available. Depending on the source, this may also be referred to as long-term point availability, fill rate, or availability of stock.
- Interval reliability: This is the probability that there will not be a stock shortage at any point within a given time frame, like a year.
- Cost minimization: This includes the price of running out of a replacement part as well as the cost of buying and stocking spares.
- Availability: This is the proportion of a system's or unit's uptime (nondowntime) when the downtime is brought on by a lack of spare components.

When a nonrepairable component fails or is removed as a preventive measure, it is quickly replaced by one from the stock (the replacement time is presumed to be minimal), and the replaced component is not repaired. It is expected that the need for spare parts follows a Poisson process, which has found widespread use in the demand for emergency components.

The ordering of spare parts is covered in a block-based maintenance setting by Chelbi and Amk T-Kadi (2001). They simultaneously optimize the length of the replenishment cycles, the time between preventive maintenance checks, and the inventory level at which fresh orders for spare parts are placed. The investigations that follow take into account both maintenance and production. Integer linear programming is used by Najid, Alaoui-Selsouli, and Mohafid (2011) to consider the multi-item capacitated lot sizing problem in conjunction with routine preventive maintenance and minimal repairs after failure. Using a Markov decision process formulation, Borrero and Akhavan-Tabatabaei (2013) take into account a single-product workstation and incorporate maintenance expenses as well as either a cost for retaining inventory or a cost for not meeting demand while the machine is not in operation. To examine the joint optimization of an unstable manufacturing system's production, setup, and maintenance operations that produces two different types of products, Assid, Gharbi, and Hajji (2015) created a simulation model. Last but not least, Zhao, Mizutani, and Nakagawa (2015) take into account a unit processing a series of jobs with random processing times. The age replacement policy that causes jobs to be interrupted is contrasted with replacement after a set number of completed jobs and replacement following the end of the job, during which the maintenance age is attained.

Modern engineering systems and production procedures are becoming complex. As a result, managing system reliability in contemporary dynamic operating contexts becomes difficult (Lee, Ghaffari, and Elmeligy, 2011). Contrary to more conventional approaches that focus on time-based maintenance (TBM), *CBM* is a prominent strategy for the scheduling of maintenance interventions in this complex system (Jardine, Lin, and Banjevic, 2006).

Even in modern industrial systems, the competitiveness of industrial enterprises could gain significantly from an ideal CBM strategy (e.g., increasing system availability and consequent benefits for safety management (Harrou, Sun, and Madakyaru, 2016), reducing maintenance costs (Wiboonrat, 2018), and increasing product quality (Harrou, Sun, and Khadraoui, 2016)). Given that CBM is a well-established study area, the existence of earlier literature analyses is anticipated. In one of the studies of global CBM industrial practices up to 1994 (Martint, 1994), fundamental concepts including the requirement for scheduled maintenance, the purpose of CM, and the distinction between diagnosis and prediction were all introduced. More recent developments in data integration for CBM (Campos, 2009) and statistical data-driven approaches for the estimation of a system's remaining useful life (RUL) (Si *et al.*, 2011) are also examined in other studies, in addition to the methods typically used to monitor the health of mechanical systems and the decision models employed (A. H. C. Tsang, 1995).

At the moment, CBM is a well-liked predictive maintenance strategy. In recent decades, CBM techniques and procedures have been progressively enhanced. Due to the inherent benefits of using information retrieved from numerous sensors, sensor fusion techniques are increasingly and widely used. There has been a lot of focus on a number of approaches, including vibration, temperature, acoustic emissions, ultrasonic, oil debris, lubricant condition, chip detectors, and time/stress assessments. Since they are so good at representing machine performance, for instance, vibration signature analysis, oil analysis, and acoustic emissions, they have long been used successfully for prognostics. Three primary categories can be used to categorize current prognostic methods, which are as follows:

Model-based approach: This approach demands in-depth familiarity with the physical interactions and traits of each connected component in a system. The actual operating state is determined by measurements, and the expected operating state is derived from the values of the characteristics obtained from the physical model. The model-based method is typically tough, as the relationships and traits of all associated components in a system and its environment are frequently too complex to be represented by a model with an acceptable level of accuracy. Additionally, it is possible that the values of some process parameters or components are not readily available.

Data-driven approach: This approach demands a lot of historical information that depicts both "regular" and "faulty" operations. Instead of using prior process knowledge, it merely takes measurement data from the process itself to generate behavioral models. In this method, pattern recognition algorithms are frequently employed. The results of data analysis can be interpreted using general process knowledge, and then qualitative methods like fuzzy logic and artificial intelligence can be employed for decision-making to enable failure avoidance.

Hybrid approach: This approach combines sensor-based information with model-based information and employs both model-driven and data-driven methodologies to create prognostic outcomes that are more accurate and dependable.

The first industrial revolution began with corrective maintenance, and the second industrial revolution started with scheduled maintenance during the Ford automotive era. The two primary phenomena of the third industrial revolution are automation and computerization. With the aid of early automation and computerization, this has led to the usage of productive maintenance, also known as total productive maintenance or reliability-centered maintenance. The most advanced type of maintenance is "predictive maintenance," which was made possible by the relatively new Industry 4.0 concept, which fully utilized the benefits of cyber systems, cloud storage, and the Internet of Things. There is also an inverse relationship between the amount of maintenance and its "factor" OEE, as shown in Table 1.2 below. The overall effectiveness of the equipment increases as maintenance becomes more sophisticated.

The industry has a significant opportunity to deploy novel solutions to enhance operations and maintenance (O&M) practice given the improvement in instrumentation technologies, analytical software, and mathematical modeling. This has increased the level of optimism in several industries that still rely on traditional O&M procedures, encouraging those sectors to take advantage of multiple chances to lower the commercial risks connected with plant operations. Risk reduction and value creation are defined by technical circumstances and the safety integrity of operational plants or assets.

It suggests that a key component of risk mitigation and value creation initiatives is the operator's capacity to recognize equipment or system failures before any undesirable event or incident. In theory, this capability depends heavily on the technical information collected from the technical

Table 1.2 Correlation of the industrial revolution and maintenance (Coleman, 1956).

Industry revolution	**Industry 1.0**	**Industry 2.0**	**Industry 3.0**	**Industry 4.0**
Characteristics of the industrial revolution	Mechanization, steam power, weaving loom	Mass production, assembly lines, electrical energy	Automation, computers, electronics	Cyber-physical systems, IoT, networks, cloud, BDA
Type of maintenance	Reactive maintenance	Planned maintenance	Productive maintenance	Predictive maintenance
Inspection	Visual inspection	Instrumental inspection	Sensor monitoring	Predictive analysis
OEE (overall equipment effectiveness)	<50%	50–75%	75–90%	>90%
Maintenance team reinforcement	Trained craftsmen	Inspectors	Reliability engineers	Data scientists

apparatus and systems, as well as the operator's decision-support configuration. In order to gather the necessary technical data, critical systems and equipment must be properly instrumented. Analytical software with incorporated mathematical models is essential for the decision-making process. The core fundamentals of the O&M intervention process, which attempts to maintain or restore systems or equipment to a given condition so that the plant or asset conforms with a specific level of performance, are really presented in this explanation (see Figure 1.9). The technical tools in use and the analytical software and tools offer the essential engineering foundation for keeping track of the health of any specific asset's systems and equipment. The outcomes of this condition monitoring procedure are used as inputs by the asset operator or plant decision platforms and processes to make diagnostic or prognostic choices. Work orders are issued for the O&M crew whenever a fault or failure is about to occur. The most fundamental principle of the CBM approach is illustrated by this O&M intervention method. The widely adopted ideas of "e-maintenance" and "intelligent maintenance systems" can take advantage of the CBM platform's availability and manifest as sophisticated applications utilizing information and computer technology, durable technical infrastructures, and electronic devices and data acquisition technologies.

Studies have acknowledged that condition monitoring is not always accurate. These investigations take into account both the scenario in which the defective state is not detected and the scenario in which the inspection

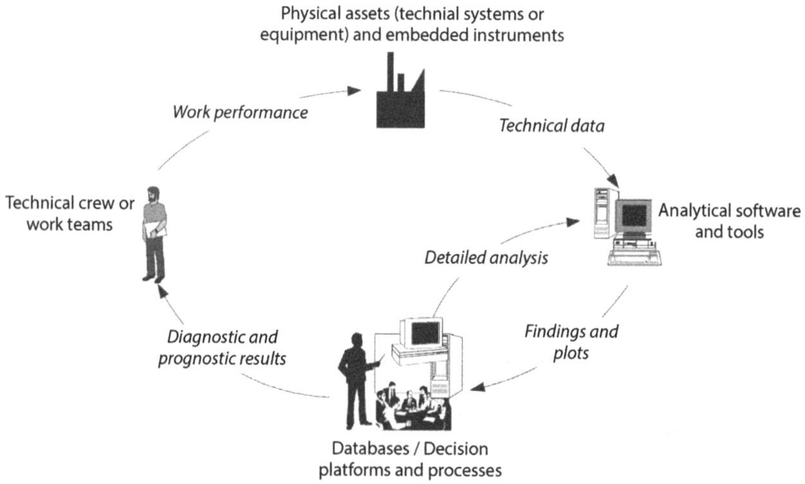

Figure 1.9 Basic O&M intervention process to retain or restore technical systems and equipment of an industrial asset in an acceptable technical condition.

inadvertently detects the defective state. Some publications take into account the proportional hazards model, in which the failure rate depends on the equipment's age and condition. Maintenance actions like overhaul and repair can be thought of as being equivalent to a replacement if it is reasonable to assume that they also restore equipment to the condition it was in before the repair or overhaul was performed. As a result, the following models can frequently be utilized to examine overhaul/repair issues in practice because this assumption is frequently a fair one.

1.3 Repair

In this section, we just provide a generic overview of various repair fundamentals with a specific introduction to imperfect repair. However, a detailed mathematical understanding of the imperfect repair process is provided in the second chapter.

The failure modes producing an asset's dysfunction determine the extent of repair that is required. Equipment failure can be classified into two categories: intermittent and extended failure. Intermittent failure lasts for a short time and occurs under certain conditions intermittently. Extended failure needs a certain corrective action or else it continues. Extended failure is further divided into two parts: partial and complete failure. Partial failures result in partial loss of function. Systems work but at a reduced capacity. Complete failures are the result of a total loss of function. The resource is totally broken and cannot be utilized until it is fixed. The further classification of complete and partial failures is mentioned in Figure 1.10.

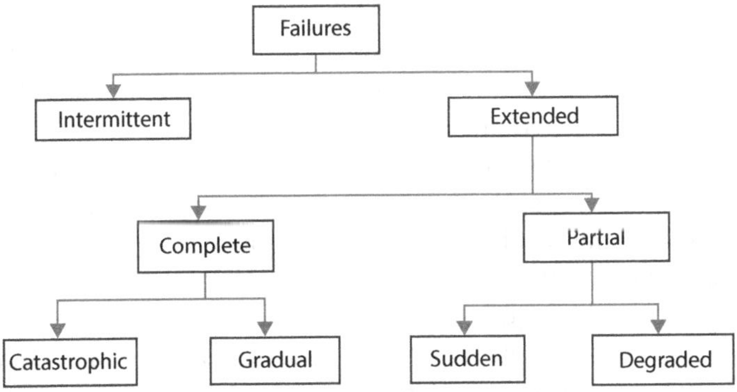

Figure 1.10 Types of failure.

When an item breaks down, it can be exceedingly expensive to repair it as well as to delay or cease production. Some failures are the result of mistakes made by people, unknown factors, or the normal wear and use of assets over many years of use. Many of these conditions are avoidable with preventive maintenance.

When a failure occurs, a system capable of being repaired rather than completely replaced can be said to be repairable. A system that cannot be repaired and must be replaced after malfunctioning is said to be nonrepairable. Many systems in the real world, including those in cars, planes, computers, and air conditioners, are repairable. The reliability of the repairable system made up of nonrepairable components will undoubtedly increase as a result of the increase in the reliability of the nonrepairable components.

Traditional reliability life or accelerated test data analysis, nonparametric or parametric, is based on a truly random sample drawn from a single population and independent and identically distributed (i.i.d.) assumptions on the reliability data obtained from the testing/fielded units. This i.i.d. assumption may also be valid, intuitively, on the first failure of several identical units, coming from the same design and manufacturing process, fielded in a specified or assumed to be in an identical environment. The life data of such items usually consist of an item's single failure (or very first failure for reparable items) and the times of some items still surviving, referred to as censoring or suspension. The reliability literature is vast to cover such aspects in reliability data analysis where the failure times are modeled by appropriate life distributions.

However, in repairable systems, one generally has times of successive failures of a single system, often violating the i.i.d. assumption. Hence, it is not surprising that statistical methods required for repairable systems differ from those needed in the reliability analysis of nonrepairable items. In order to address the reliability characteristics of complex repairable systems, a process rather than a distribution is often used. For a repairable system, time to next failure depends on both the life distribution (the probability distribution of the time to first failure) and the impact of maintenance actions performed after the first occurrence of a failure. The most popular process model is the power law process (PLP). This model is popular for several reasons. For instance, it has a very practical foundation in terms of minimal repair. Second, if the time to first failure follows the Weibull distribution, then the power law model repair governs each succeeding failure and adequately models the minimal repair phenomenon. In other words, the Weibull distribution addresses the very first failure, and the PLP addresses each succeeding failure for a repairable system. From this viewpoint, the PLP can be regarded as an extension of the Weibull distribution and a generalization of the Poisson process. Besides, the PLP is generally computationally easy in providing useful and practical solutions, which have been usually comprehended and accepted by the management for many real-world applications.

Minimal repair: Minimal repair restores the system to an "as bad as old condition" (ABAO). The assumption of minimal repair leads to the non-homogeneous Poisson process (NHPP). The NHPP is often a good model for repairable systems because it can model systems that are deteriorating or improving.

Renewal (or perfect) repair: Perfect repair restores the system to an "as good as new condition" (AGAN). If every repair is a renewal repair, then the times between failures are independent and identically distributed. Mathematically perfect repair is modeled using a renewal process (RP).

Imperfect repair: A failed unit is either repaired to a like-new condition or to a bad as old condition (i.e., a condition identical to that of a functioning unit of the same age). Specifically, under the imperfect repair model, a failed unit is given a perfect repair with probability p and an imperfect repair with probability $q = 1 - p$. In special cases, if $p = 0$, we have an NHPP, and if $p = 1$, we have a renewal process. A better understanding of these repair processes is shown below in Figure 1.11.

The following selected terms pertaining to the repairable systems are useful:

1. Point process
2. Counting random variable

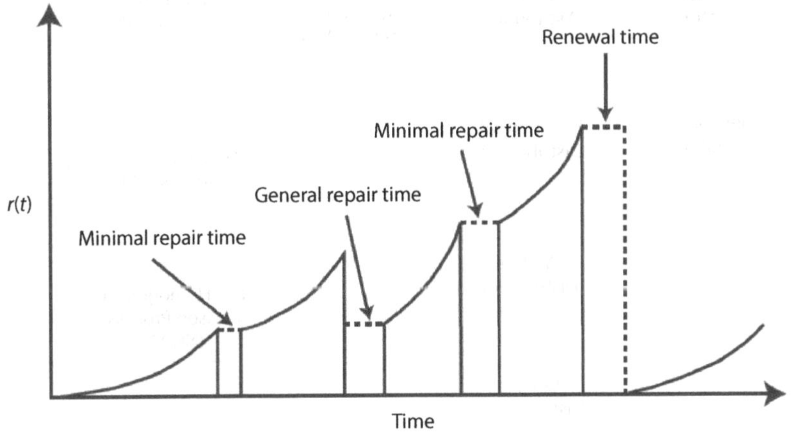

Figure 1.11 Minimal and general repair (Jardine and Tsand, 2005).

3. Mean function of the point process
4. Rate of occurrence of failures
5. Intensity function

All these terminologies are covered in detail in Chapter 2. Figure 1.12 summarizes the techniques in vogue for reliability analysis for both repairable and nonrepairable items, respectively.

Graphical methods for displaying data from repairable systems can be used to gain insight into the data and also to select a reasonable model. The first question one should ask when analyzing data from repairable systems is whether there is a trend in the times between failures. That is, are the times between failure getting longer? Or are they getting shorter? A simple, but powerful, graphical method is to plot the (global) failure time t_i along the

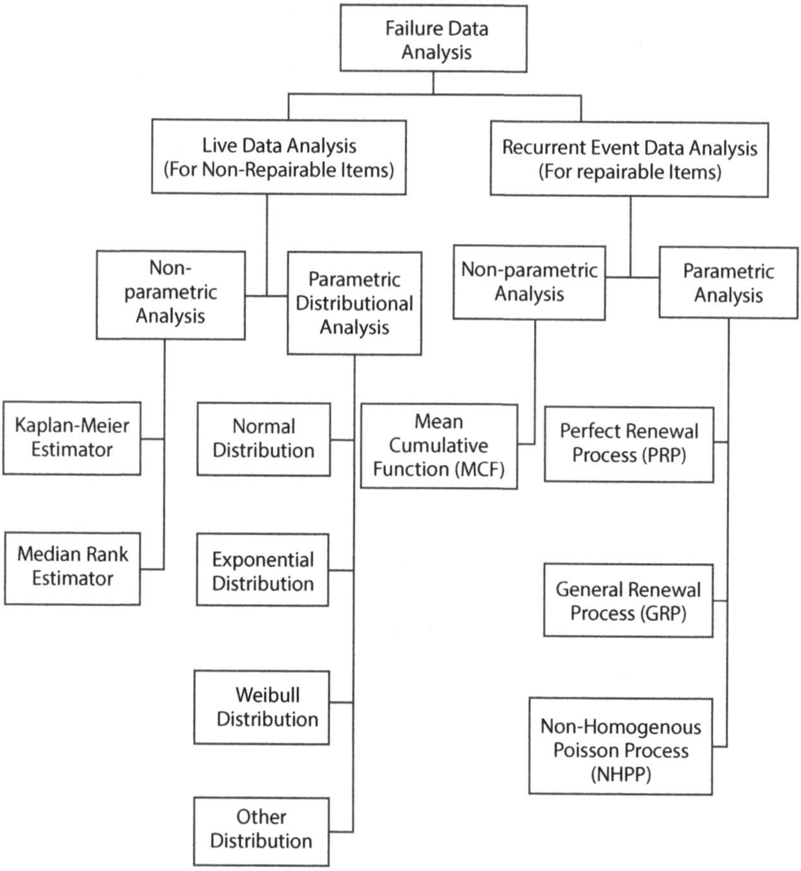

Figure 1.12 Various techniques for reliability analysis (Rai, Chaturvedi, and Bolia, 2020).

horizontal axis, and the cumulative number of failures through time t_i, $N(t)$, on the vertical axis. It is widely believed that Duane plots should be roughly linear if the power law process governs the failure times of the repairable system. Total time on test (TTT) (Bo Bergman, 2008) plots were developed for selecting a model for the lifetime distribution of a repairable system.

The homogeneous Poisson process (HPP) is the simplest statistical model for describing the occurrence of failures of a repairable system. The HPP, however, carries with it some strong assumptions that in practice are often clearly violated. The HPP implies that the system does not age in global time (failure time since the initial start-up), that is, the system does not deteriorate and it does not exhibit reliability improvement. The HPP also implies that the system does not wear out in local time (failure time since the previous failure), that is, during the times between failures. If either of these assumptions does not hold, then the HPP is not the appropriate model. Graphical methods are useful for determining the suitability of the HPP. If the plot of $N(\mathrm{t})/t$ versus t is linear, or if the Duane plot has a zero slope, then the HPP may be an appropriate model. It would then be useful to look at the times between failures to see if an exponential distribution can model them. If they can, then the HPP is appropriate; if they cannot, then a renewal process with a distribution besides the exponential would be appropriate.

A nonhomogeneous Poisson process is similar to an ordinary Poisson process, except that the average rate of arrivals is allowed to vary with time. Many applications that generate random points in time are modeled more faithfully with such nonhomogeneous processes. A homogeneous Poisson process has a constant intensity function, while a nonhomogeneous Poisson process can have a not constant intensity function.

Several statistical tests have been proposed for testing the null hypothesis that the process is an HPP against the alternative that there is a monotonic trend. Two types of tests for this purpose include the Laplace test, one of the earliest statistical tests of any kind, and the MIL-HDBK-189 test.

For a company or a competing manufacturer, the common concerns can be (Tobias, 2012; Nelson, 2003):

- The number of repairs, on average, for all systems at a specified operational time.
- Expected time to first repair, subsequent repairs, etc. Trends in repair rate or costs whether increasing, decreasing, or substantially constant.
- How to take decisions on burn-in requirements and maintenance or retirement.

- Is burn-in beneficial? How long and cost-effective would it be?
- How to compare different versions, designs, or performance of systems operating in different environments/regions, etc.

Each of the above questions and many more can be answered by the mean cumulative function (MCF) method based on the nonparametric graphical approach. It was observed that the MCF-based nonparametric method is a better choice to analyze the nature of fleet and system recurrence rates through the operational life of a product.

To simulate difficulties with system maintenance that can be fixed, the virtual age idea has been created. The idea of virtual age was introduced by Malik (1979) and Kijima (1989), respectively, when modeling repairable systems. Jiang (2001) proposed a repair limit for determining what maintenance action should be taken at the time of a maintenance issue. They did this by applying the virtual age idea. Figure 1.13 provides an illustration of the strategy. Here, it can be seen that the virtual age is on the *x*-axis and the real age (run time) of an object is on the *y*-axis at the top of the figure.

As a result, following the initial maintenance intervention, when the equipment is of age X1, it is of age V1 after the maintenance action. The virtual age of the item will be V2 once it has been in use for a further length of time X2. It is important to answer the following question: Which course of action should be followed when a maintenance intervention is required? Should the equipment undergo minimal, general, or complete renewal? How much money should be spent on each type of repair? Using the cost on the *y*-axis in the bottom half of the picture, we can see the repair limit notion in Figure 1.13 and answer this query. A minimal repair is conducted if the maintenance cost estimate is between zero and the limit C0. A general repair is carried out if the cost is between C0 and the cost boundary. A general repair rather than a total replacement should be done since, as shown in the figure, the cost of the first maintenance intervention is C1. As a result, the equipment's health is better than it was before the repair. A complete renewal should occur if the cost is higher than the boundary curve. Similar to this, a complete renewal is required if the equipment's running time crosses the line for preventive replacement.

The landmark papers by Kijima (1989) and Kijima M. (1986) modeled imperfect repair using the generalized renewal process (GRP) with the idea of the "virtual age." The paper spurred tremendous growth in the literature on imperfect maintenance as a consequence of which the GRP approach is well established. One reason for its ubiquitous popularity is its general approach to model repairable systems that can incorporate all of the five

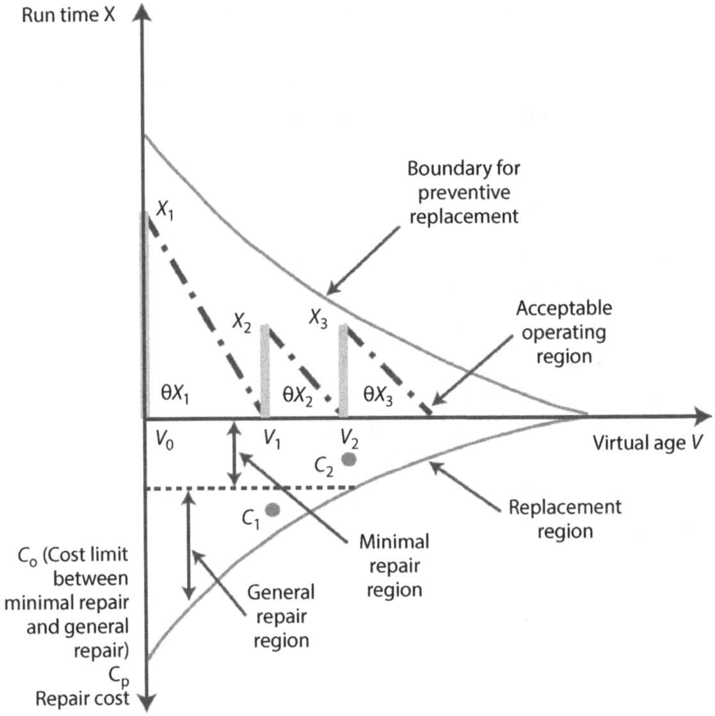

Figure 1.13 Optimizing minimal and general repair decisions (Jardine and Tsand, 2005).

after repair states in the same mathematical framework. The generalized renewal process framework is based on two broad approaches for imperfect repair: arithmetic reduction of age (ARA) and arithmetic reduction of intensity (ARI). In ARA models, the effect of repair is expressed by assuming a reduction in the actual age of the system and attaining an age termed as virtual age. ARA-based Kijima's virtual age models are the most widely cited and effective models in the literature. In the ARI approach, the repair effect is considered by the change induced on the failure intensity before and after failure. The parameters for the GRP model are also estimated using maximum likelihood estimation (MLE). Again, a detailed coverage in this regard has been provided in Chapter 1.

1.4 Overhaul

Overhaul is a detailed examination of all components and subsystems of the existing system and is a combination of preventive, corrective, and predictive maintenance. Overhaul includes thorough examination, dismantling of the whole system, repairs or replacements of the parts if required, and functional testing of systems, subsystems, and components, followed by reassembling of the complete system.

1.4.1 Overhauling: What Is It?

It is defined as a process of general maintenance performed on a system or other industrial equipment. The goal of overhauling is to keep the system in serviceable condition. Regular checks can prevent all kinds of critical damage. Machinery overhaul is usually performed by companies offering maintenance services. The frequency of overhauling can be agreed upon, and routine maintenance is usually scheduled for once a year. A more frequent equipment check is recommended for older machines and especially larger machines involving complex mechanisms. A major overhaul on a piece of equipment can also be considered as a planned corrective maintenance activity because it is undertaken at a predetermined time, usually after the dudes of the equipment are completed.

1.4.1.1 Benefits of Overhauling

(a) Cost-Effective
While it may seem unreasonable to further invest in a system that is still in operation, overhauling of the system protects the organization from undesirable expenses in the future. Replacing a few parts here and there is without a doubt more cost-effective than buying new machinery. A sudden breakdown of a machine can cause major financial losses in the industry. Even a single machine out of order means thousands and thousands of dollars lost in potential profit. Major global industrial companies recognize timely repair as a way to minimize costs and prevent critical equipment failures.

(b) Extended Life Length
Regular maintenance extends the life length of the equipment. Especially, predictive overhauling allows potential issues to be fixed even before the failure occurs. Naturally, when a piece of industrial equipment is in use, the mechanisms tend to wear down with time. However, extensive use of machinery which has shown signs of wear-down and damage can lead to a critical breakdown. Overhauling stops the extensive damage of the machinery and increases the lengths of its life cycle.

(c) Increased Performance
After years of extended use, the performance of the machinery is nowhere near its productivity at the beginning of the life cycle. With an overhaul, the machine's performance can be restored up to 100%. Such an impressive number basically means one can enjoy the benefits of using the almost brand-new equipment without any limitations. Increased performance of the machine represents the possibility to increase the level of production, hence expanding the company's profit.

(d) Reduced Labor Costs
Replacement of the entire piece of industrial equipment is essentially more time-consuming than a small repair or replacement of just one part. Instead of replacing the entire machine with a new one, only certain parts are replaced or even repaired. Predictive maintenance reduces the number of critical "call-outs" and naturally the amount of time spent on a service call.

1.4.1.2 Factors Determining the Performance of the Overhaul

(a) Time Taken for Completing the Overhaul
Overhaul activities should be carried out in such a way that each activity carried out in the overhaul is done with optimum time and cost so that the overhaul is completed in optimum time.

(b) The Quantum of the Job Carried Out
This factor describes the extent to which the overhaul is carried out, i.e., the level of disassembly and assembly done during the overhaul, whether the disassembly takes place at the subsystem level, component level, etc.

(c) The Extent to Which the System Is Restored to as Good as New State
During the overhaul, the components are replaced or repaired based on their condition found during the inspection. So, it is important to note that not all the components of the system are replaced or repaired. So, the state of the system after the overhaul is between as bad as old and as good as new condition and the extent to which the system is restored to the as good as new state depends on the extent of the activities performed during the overhaul process.

(d) The Guaranteed Period of Failure Free Operation
After the overhaul is completed, it is important to find the extent to which the overall life of the system can assure no failure of the system during the extended life of the system.

1.4.1.3 *Overhauling of the System in Stages*

Overhauling involves several stages as shown in Figure 1.14 and these are explained in subsequent paragraphs.

(a) Inspection
First of all, the machine will be thoroughly inspected. Experienced maintenance crews perform an inspection on the overhauled machine under production conditions. It means that the machine's performance is monitored while the machine is in use. Such a procedure allows to allocate any issues and perform the troubleshooting more effectively.

(b) Disassembly
After the initial inspection, the piece of equipment should be taken apart. Disassembly is crucial for further checking and for the next steps of the overhauling process, such as repair. A skilled maintenance worker is capable of putting the machine down efficiently, indicating which parts of the equipment need to be replaced or repaired.

(c) Repair
Depending on the issue, the machine is either repaired or certain damaged parts are replaced. This step once again proves how effective overhauling is as opposed to replacing the whole piece of equipment at once. Replacement of parts might take longer than a simple repair, as the spare parts might need to be ordered from a manufacturer.

(d) Reassembly
Following the successful replacement of spare parts, reassembly of the whole mechanism is performed. Being one of the final steps, reassembly is crucial for the functioning of the equipment. A certain skill is surely needed to perform reassembly, so it is best handled by professionals.

(e) Testing
This is the final step that concludes the overhauling process. Without testing, it is naturally impossible to identify if the performed repair was effective. During testing, the retrofit is either proclaimed successful or—less frequently—the process goes back to the starting point (inspection).

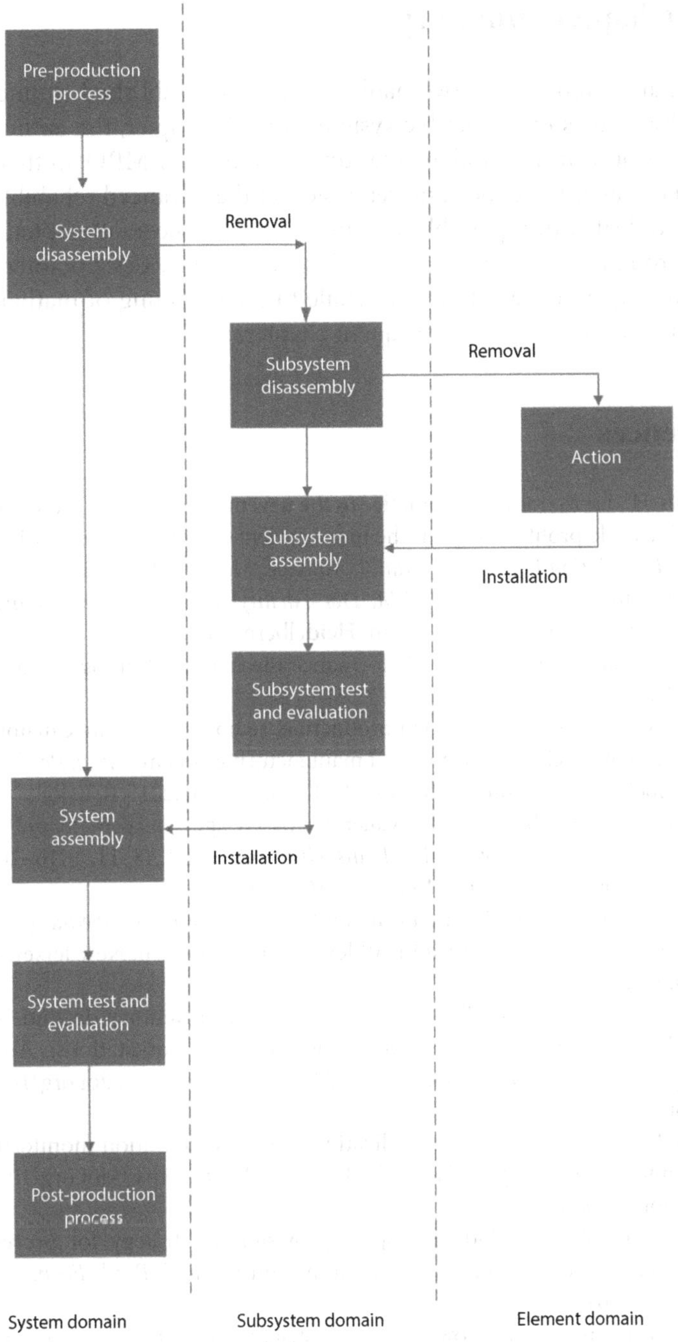

Figure 1.14 Major overhaul flowchart.

1.5 Chapter Summary

Maintenance, repair, and overhaul facilities are established to undertake reliability studies on repairable systems. In this chapter, the authors provide a preview of the fundamental understanding of MRO so that it will be easier to assimilate the comprehension of the advanced reliability techniques to deal with repairable systems, which has been attempted in this book through subsequent chapters. This chapter provides a pointer on the basics of imperfect repair, and a detailed understanding of mathematical concepts is covered in the subsequent chapters.

References

Ahlmann, H., From traditional practice to the new understanding: The significance of life cycle profit concept in the management of industrial enterprises. *Proc. Int. Foundation Res. Maint., Maint. Manage. Model.*, 2002, May, 6–7

Altmannshoffer, R., *Industrielles FM, Der Facility Manager (In German)*, April Issue, pp. 12–30, Springer, Berlin, Heidelberg, 2006.

Andersen, B. and Fagerhaug, T., Performance measurement of logistic processes, 2007.

Assid, M., Gharbi, A., Hajji, A., Joint production, setup and preventive maintenance policies of unreliable two-product manufacturing systems. *Int. J. Prod. Res.*, 53, 15, 4668–83, 2015, https://doi.org/10.1080/00207543.2015.1030468.

Bajestani, M.A. and Banjevic, D., Calendar-based age replacement policy with dependent renewal cycles. *IIE Trans. (Inst. Ind. Eng.)*, 48, 11, 1016–26, 2016, https://doi.org/10.1080/0740817X.2016.1163444.

Bergman, B. and Klefsjo, B., Total time on test plots, in: *Encyclopedia of Statistics in Quality and Reliability*, John Wiley & Sons, Hoboken, New Jersey, United States, 2008.

Borrero, J.S. and Akhavan-Tabatabaei, R., Time and inventory dependent optimal maintenance policies for single machine workstations: An MDP approach. *Eur. J. Oper. Res.*, 228, 3, 545–55, 2013, https://doi.org/10.1016/j.ejor.2013.02.011.

Campos, J., Development in the application of ICT in condition monitoring and maintenance. *Comput. Ind.*, 60, 1, 1–20, 2009, https://doi.org/10.1016/j.compind.2008.09.007.

Chelbi, A. and Amk T-Kadi, D., Spare provisioning strategy for preventively replaced systems subjected to random failure. *Int. J. Prod. Econ.*, 74, 1–3, 183–9, 2001.

Coleman, D.C., Industrial growth and industrial revolutions. *New Ser.*, 23, 1–22, 1956.

Dekker, R. and Plasmeijer, R.P., Multi-parameter maintenance optimisation via the marginal cost approach. *J. Oper. Res. Soc.*, 52, 2, 188–97, 2001, https://doi.org/10.1057/palgrave.jors.2601072.

Duffuaa, S.O. and Al-Sultan, K.S., Mathematical programming approaches for the management of maintenance planning and scheduling. *J. Qual. Maint. Eng.*, 3, 3, 163–176, 1997.

Finkelstein, M., Shafiee, M., Kotchap, A.N., Classical optimal replacement strategies revisited. *IEEE Trans. Reliab.*, 65, 2, 540–46, 2016. https://doi.org/10.1109/TR.2016.2515591.

Geraerds, W.M.J., *The cost of downtime for maintenance: Preliminary considerations*, University of Technology, Department of industrial engineering & management science, Eindhoven, Netherlands, 1983.

Harrou, F., Sun, Y., Khadraoui, S., Amalgamation of anomaly-detection indices for enhanced process monitoring. *J. Loss Prev. Process Ind.*, 40, March, 365–77, 2016. https://doi.org/10.1016/j.jlp.2016.01.024.

Harrou, F., Sun, Y., Madakyaru, M., Kullback-Leibler distance-based enhanced detection of incipient anomalies. *J. Loss Prev. Process Ind.*, 44, November, 73–87, 2016. https://doi.org/10.1016/j.jlp.2016.08.020.

He, K., Maillart, L.M., Prokopyev, O.A., Optimal planning of unpunctual preventive maintenance. *IISE Trans.*, 49, 2, 127–43, 2017. https://doi.org/10.1080/0740817X.2016.1224959.

Jardine, A.K.S., Lin, D., Banjevic, D., A review on machinery diagnostics and prognostics implementing condition-based maintenance. *Mech. Syst. Sig. Process.*, 20, 7, 1483–1510, 2006. https://doi.org/10.1016/j.ymssp.2005.09.012.

Jardine, A.K.S. and Tsang, A.H.C., *Maintenance, replacement, and reliability: Theory and applications*, CRC Press, Boca Raton, Florida, 2005.

Jiang, X., Makis, V., Jardine, A.K.S., Optimal repair/replacement policy for a general repair model. *Adv. Appl. Probab.*, *33*, 206–22, 2001.

Kijima, M., Results for repairable systems with general repair. *J. Appl. Probab.*, 29, 89–102, 1989.

Kijima, M. and Sumita, U., A useful generalization of renewal theory: Counting processes governed by non-negative markovian increments. *J. Appl. Probab.*, 23, 1, 71–88, 1986.

Lee, J., Ghaffari, M., Elmeligy, S., Self-maintenance and engineering immune systems: Towards smarter machines and manufacturing systems. *Annu. Rev. Control*, 35, 1, 111–22, 2011. https://doi.org/10.1016/j.arcontrol.2011.03.007.

Malik, M.A.K., Reliable preventive maintenance scheduling. *AIIE Trans.*, 11, 3, 221–28, 1979. https://doi.org/10.1080/05695557908974463.

Martint, K.F., A review by discussion of condition monitoring and fault diagnosis in machine tools. *Int. J. Mach. Tools Manuf.*, 34, 89, 183–9, 1994. 0890-6955(94)E0015-B.

Murthy, D.N.P., Atrens, A., Eccleston, J.A., Strategic maintenance management. *J. Qual. Maint. Eng.*, 8, 4, 287–305, 2002. https://doi.org/10.1108/13552510210448504.

Najib, N.M., Alaoui-Selsouli, M., Mohafid, A., An integrated production and maintenance planning model with time windows and shortage cost. *Int. J. Prod. Res.*, 49, 8, 2265–83, 2011. https://doi.org/10.1080/00207541003620386.

Nakajima, S., *Introduction to TPM: Total Productive Maintenance (Preventative Maintenance Series)*, Productivity Press, New York, United States, 1988.

Nelson, W.B., Recurrent events data analysis for product repairs, disease recurrences and other applications. *SIAM*, 2003.

Rai, R.N., Chaturvedi, S.K., Bolia, N., *Repairable systems reliability analysis: A comprehensive framework*, John Wiley & Sons, Hoboken, New Jersey, U.S., 2020.

Si, X.S., Wang, W., Hu, C.H., Zhou, D.H., Remaining useful life estimation - A review on the statistical data driven approaches. *Eur. J. Oper. Res.*, 213, 1, 1–14, 2011, Elsevier B.V, https://doi.org/10.1016/j.ejor.2010.11.018.

Sidibe, I.B., Khatab, A., Diallo, C., Kassambara, A., Preventive maintenance optimization for a stochastically degrading system with a random initial age. *Reliab. Eng. Syst. Saf.*, 159, March, 255–63, 2017, https://doi.org/10.1016/j.ress.2016.11.018.

Tobias, P.A. and Trindade, D.C., *Applied reliability*, Chapman & Hall/CRC, Boca Raton, Florida, 2012.

Tsang, A.H.C., Condition-based maintenance 3 condition-based maintenance: Tools and decision making. *J. Qual. Maint. Eng.*, 1, 3, 3–17, 1995.

Tsang, A.H.C., Strategic dimensions of maintenance management. *J. Qual. Maint. Eng.*, 8, 1, 7–39, 2002. https://doi.org/10.1108/13552510210420577.

Wiboonrat, M., Developing diagnostics and prognostics of data center systems implementing with condition-based maintenance. *IECON 2018-44th Annual Conference of the IEEE Industrial Electronics Society*, pp. 4901–4906, 2018.

Wireman, T., *Developing performance indicators for managing maintenance*, Industrial Press, Inc, New York, 1998.

Wireman, T., *How to calculate return on investment for maintenance improvement projects*, Iron and steel engineer: International technology for world competition, Pittsburgh, 2007.

Zhao, X., Mizutani, S., Nakagawa, T., Which is better for replacement policies with continuous or discrete scheduled times? *Eur. J. Oper. Res.*, 242, 2, 477–86, 2015. https://doi.org/10.1016/j.ejor.2014.11.018.

2

Repairable Systems: An Overview

After introducing the readers to a preview on the fundamentals of MRO in the first chapter, we now undertake a detailed overview of the basic features of repairable systems. The understanding of the concepts of repairable systems as provided in this chapter will help the readers and the research fraternity in a better comprehension of the advanced techniques utilized for modeling and reliability analysis of repairable systems which forms the main theme of this book.

Maintenance modeling and analysis of repairable systems (Rigdon and Basu, 2000) have gained large attention from researchers in the last few decades because of their desired applications in large-scale industries. With the advancement in technology, the competition to survive in the market is increasing manifold for both the manufacturing and service industries. Now, industries seek for optimal reliability solutions to improve the overall performance of their assets. Most of the industries which deal with large, complex, and critical systems, such as automobile, aerospace, and nuclear industries, provide due attention to the performance, maintenance scheduling, and reliability improvement of the systems (Jardine and Tsang, 2005). Generally, these large and complex systems are repairable and often subjected to imperfect maintenance and, hence, require extra attention in terms of reliability analysis. The industries also seek such models and maintenance policies that can provide remedial solutions to upgrade performance. For this purpose, most of the industries have their own setup of maintenance, repair, and overhaul (MRO) facilities. The MRO facilities are entrusted with the task of carrying out maintenance, repair, and overhaul of complex equipment and also conduct life data analysis for reliability and availability improvement of the system. In addition, they also review the quality of maintenance tasks for improving the performability of the systems. The content, presented in this book, is dedicated toward providing novice models and methodologies to deal with modeling and analysis of repairable systems which are subjected to imperfect maintenance in MRO facilities. We mainly focus on modeling realistic situations, analyzing the

Garima Sharma and Rajiv Nandan Rai. Advanced Techniques for Maintenance Modeling and Reliability Analysis of Repairable Systems, (37–64) © 2023 Scrivener Publishing LLC

generated failure data more scientifically, and developing methodologies and maintenance plans for improving the overall performance of repairable systems. Before moving forward into the discussion, it is essential to understand the concept of repairable systems and related maintenance actions in detail. In the following section, we briefly introduce the repairable systems and their basic terminologies related to reliability engineering.

2.1 Introduction to Repairable Systems and Its Terminologies

A system is a collection of mutually related items, assembled to perform one or more intended functions. Any system majorly consists of (i) items as the operating parts, (ii) attributes as the properties of items, and (iii) the link between items and attributes as interrelationships. A system is not only expected to perform its specified function(s) under its operating conditions and constraints but also expected to meet specified requirements, referred to as performance and attributes. The failure of a system, therefore, can be defined as an event whose occurrence results in either loss of ability to perform required function(s) or loss of ability to satisfy the specified requirements (i.e., performance and/or attributes). Regardless of the reason for the occurrence of this change, a failure causes the system to transit from a state of functioning to a state of failure or a state of unacceptable performance. For many systems, a transition to the unsatisfactory or failure state means retirement. Engineering systems of this type are known as *nonmaintained* or *nonrepairable* systems because it is impossible to restore their function ability within a reasonable time, means, and resources. For example, a missile is a nonrepairable system once launched. Other examples of nonrepairable systems include electric bulbs, batteries, and transistors. However, there is a large number of systems whose function ability can be restored by effecting certain specified tasks known as *maintenance tasks*. These tasks can be as complex as necessitating a complete overhaul or as simple as just cleaning, replacement, or adjustment. One can cite several examples of repairable systems in one's own day-to-day interactions with such systems that include but not are limited to automobiles, computers, aircrafts, and industrial machineries. For instance, a laptop, not connected to an electrical power supply, may fail to start if its battery is dead. In this case, replacing the battery—a nonmaintained item—with a new one may solve the problem. A television set is another example of a repairable system, which upon failure can be restored to a satisfactory condition by simply replacing either the failed resistor or transistor or even a circuit board if that is the cause or by adjusting the sweep or synchronization settings.

Thus, the system is called a repairable system if it can be restored to an operating condition by any method other than replacing the whole system (Rigdon and Basu, 2000; H. E. Ascher, 2008). Most complex systems, such as automobiles, communication systems, aircraft, engine controllers, printers, medical diagnostics systems, helicopters, and train locomotives, are repaired once they fail. In fact, when a system enters into the utilization process, it is exposed to three different performance-influencing factors, viz., operation, maintenance, and logistics, which should be strategically managed in accordance with the business plans of the owners. It is often of considerable interest to determine the reliability and other performance characteristics under these conditions. The areas of interest may include assessing the expected number of failures during the warranty period, maintaining minimum reliability for an interval, addressing the rate of wear out, determining when to replace or overhaul a system, and minimizing its life cycle costs (Rai, Chaturvedi, and Bolia, 2020).

The reliability modeling and analysis of repairable systems have always been a challenging task for the industries (Lindqvist, 2006). In this section, we summarize some important terminologies used to understand and work with repairable systems' reliability analysis. However, before that, it is very essential to understand some basic terminologies related to reliability engineering.

2.1.1 Basics of Reliability Engineering

Reliability engineering is a discipline of engineering for applying scientific know-how to a system, component, or process to ensure that without any failure, it performs its intended operation in a given time period and specified environment. Mathematically, the reliability function $R(t)$ is defined as the probability that the system will carry out its mission through time t. In other words, $R(t)$ evaluated at t is the probability at time t that the failure time is beyond time t. Thus, $R(t)$ of a random variable (which can take any positive value in this case) T can be expressed as (Rigdon and Basu, 2000):

$$R(t) = P(T > t) = 1 - P(T \leq t) \tag{2.1}$$

Cumulative distribution function (CDF): The CDF is the failure probability of a system at time t which is defined as:

$$F(t) = P(T \leq t) = 1 - P(T > t) = 1 - R(t) \tag{2.2}$$

$$F(t) = 0 \; for \; t < 0$$

Probability density function (PDF): If the derivative of CFD exists, the PDF is defined as the derivative of CDF. That is,

$$f(t) = \frac{d}{dt} F(t) = -\frac{d}{dt} R(t); \tag{2.3}$$

$$\int_0^\infty f(t)dt = 1 \tag{2.4}$$

Another way to express the PDF is through the limit

$$f(t) = \lim_{\Delta t \to 0} \frac{F(t + \Delta t) - F(t)}{\Delta t} = \lim_{\Delta t \to 0} \frac{P(t < T \leq t + \Delta t)}{\Delta t} \tag{2.5}$$

Failure rate: It is the frequency by which a system or component fails. It is expressed as failures per unit of time. Computing the failure rate over a small interval of time gives the hazard function (which is also termed as hazard rate), $h(t)$. This turns out to be the instantaneous (conditional) failure rate or instantaneous hazard rate as Δt tends to zero.

$$\begin{aligned} h(t) &= \lim_{\Delta t \to 0} \frac{P(t < T \leq t + \Delta t | T > t)}{\Delta t} \, or, \\ h(t) & \lim_{\Delta t \to 0} \frac{R(t) - R(t + \Delta t)}{\Delta t * R(t)} \end{aligned} \tag{2.6}$$

This is the limit of the probability that a system fails in a small interval given that it survived to the beginning of the interval. In other words, hazard function is the conditional probability that one and only one failure will occur in a small interval, divided by the length of the interval. The hazard function also can be written as:

$$h(t) = \frac{f(t)}{R(t)} = -\frac{dR(t)}{dt} \times \frac{1}{R(t)}$$
$$h(t)dt = -\frac{1}{R(t)} dR(t) \tag{2.7}$$

Integrating both the sides of Eq. (2.7) from 0 to t (change the variable from t to y).

$$-\int_0^t h(y)dy = \int_0^t \frac{1}{\mathrm{R(y)}} dR(y)$$

$$-\int_0^t h(y)dy = log|R(y)|\,; y = t and y = 0$$

$$-\int_0^t h(y)dy = log\ R(t) - log\ R(0)$$

$$-\int_0^t h(y)dy = log\ R(t) - log\ 1$$

$$-\int_0^t h(y)dy = log\ R(t)$$

Taking exponential both sides

$$R(t) = exp\left(-\int_0^t h(y)dy\right) \tag{2.8}$$

Thus,

$$F(t) = 1 - exp\left(-\int_0^t h(y)dy\right) \tag{2.9}$$

$$f(t) = h(t) * exp\left(-\int_0^t h(y)dy\right) \tag{2.10}$$

The above basic terminologies are mainly applicable to nonrepairable systems that get discarded after failure and follow the independently and identically distributed (i.i.d.) assumption, whereas a repairable system does not (Rolnitzky, 2011). This book primarily deals with the modeling and analysis of repairable systems, and the basic terminologies of repairable systems have been discussed in the subsequent section.

2.1.2 Terminologies for Repairable Systems

Repairable systems are generally more expensive than nonrepairable systems since it includes maintenance cost and repair cost which again includes labor cost for removing the broken part, actual repair cost, shipping cost, etc. On the other hand, nonrepairable systems can be discarded upon failure and replaced with new ones without any additional cost (Reliawiki, 2015).

Traditionally, in the literature, the point process theory is used to model the failure times of repairable systems. In addition, the occurrence of failures in repairable systems is considered a point on the time axis and is neither independent nor identically distributed. Some basic terminologies related to repairable systems are described as follows (Rigdon and Basu, 2000):

Point process: It is a stochastic model (stochastic models possess some inherent randomness) that describes the occurrence of events in time.

Counting random variable: Let us consider $N(t)$ as a random variable that denotes the number of failures in the interval $[0, t]$. When N has as its argument an interval, such as $N(a, b]$, the result is the number of failures in that interval. Here, N is termed as a counting random variable. Thus, the number of failures in the interval $(a, b]$ is defined as:

$$N(a, b] = N(b) - N(a) \tag{2.11}$$

Mean function of the point process: It is defined to be the expectations as follows:

$$\Lambda(t) = E(N(t)) \tag{2.12}$$

$\Lambda(t)$ is the expected number of failures in time t.

Rate of occurrence of failures (ROCOF): If $\Lambda(t)$ is differentiable, ROCOF is defined as:

$$u(t) = \frac{d}{dt}\Lambda(t) \tag{2.13}$$

Intensity function: It is the probability of failure in a small interval of time divided by the length of the interval.

$$u(t) = \lim_{\Delta t \to 0} \frac{P(N(t, t + \Delta t] \geq 1}{\Delta t} \tag{2.14}$$

Comparing $u(t)$ with $h(t)$, $h(t)$ is the limit of conditional probability, whereas $u(t)$ is not. Hence, $u(t)$ is the unconditional probability of a failure (not necessarily the first one) in a small interval divided by the length of the interval. The shape of the bathtub curve (Figure 2.1) of intensity function for repairable systems is similar to that of nonrepairable systems with different interpretations. It indicates that the system will experience many failures in its early operational life, trialed by a time with constant intensity function. Finally, as the system ages, the failures will be more frequent. Various maintenance actions are performed on repairable systems to restore them to an operating condition or to prevent them from failures. In the next section, the same is discussed in detail.

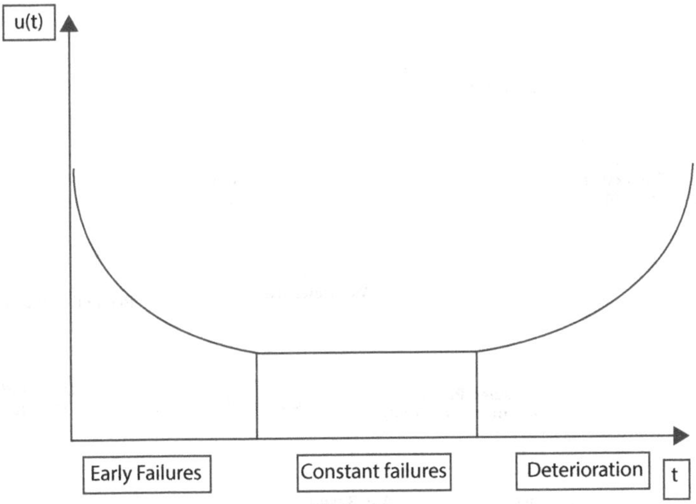

Figure 2.1 The bathtub curve for repairable systems.

2.2 Maintenance Actions on Repairable Systems

Though we have briefly introduced this topic in our previous chapter, we supplement the understanding of the maintenance concepts with supported and concerned mathematical expressions in this section.

Maintenance actions play an imperative role in the repairable system's total technical life (TTL). It affects the system's overall reliability, availability, downtime, cost of operation, etc. While dealing with repairable systems, it is very essential to understand various maintenance actions through which a repairable system undergoes.

Typically, maintenance actions are divided into two major categories (Figure 2.2), namely, reactive or corrective maintenance (CM) and proactive maintenance.

2.2.1 Reactive or Corrective Maintenance (CM)

Corrective maintenance consists of some action(s) performed on a repairable system to restore it from a failure state to an operating condition. Corrective maintenance generally involves repairing the component which is responsible for system failure. It is an unplanned maintenance action that aims to render the system from failure to an operational state within the shortest possible time. Corrective maintenance is done generally in the following three steps (Figure 2.3):

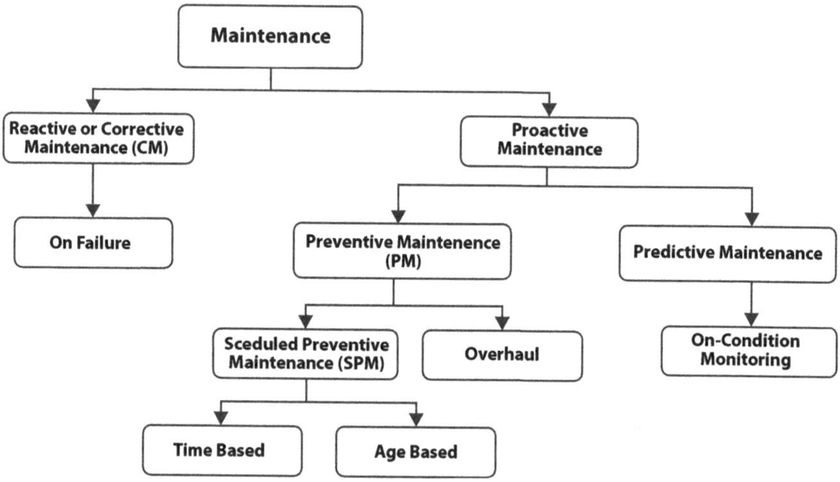

Figure 2.2 Types of maintenance actions.

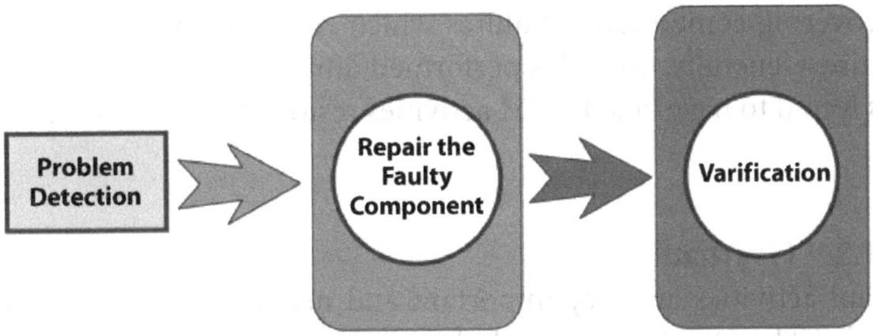

Figure 2.3 Steps for corrective maintenance.

Problem detection: The first step is to detect the problem or locate the failed component which caused the system failure.

Repairing the faulty component(s): The next step is to perform the required repair actions on the faulty component of the system. The action could be either repairing or replacing the faulty component.

Verification: Once the component(s) is (are) repaired, the action should be verified by checking whether the system is operating successfully or not.

2.2.2 Proactive Maintenance

Proactive maintenance (Ebeling, 2004) is done to prevent the system from failures. It is again divided into two categories: preventive maintenance (PM) and predictive maintenance (Yahyatabar and Najafi, 2018). Let us discuss them one by one.

2.2.2.1 Preventive Maintenance

Preventive maintenance (PM) is done when the system is in operating condition. In this practice, the components are checked or replaced, before they fail, to enhance the performability of the system. It includes scheduled PM (SPM) and periodic overhauls.

2.2.2.1.1 Scheduled Preventive Maintenance (SPM)

Scheduled preventive maintenance is done in a fixed time interval and includes some regular activities like oil filtering, lubrication, and functional testing. It can be both time- and age-based. The scheduling of these maintenance actions could be based on the system's historical data or the wear-out profile of the components. These inspections are important for making the operations smooth in upcoming working hours and also help in discovering some hidden failures which can cause a system failure in the future. Generally, no CM is performed during SPM unless any component is found to have failed. SPM activities restore the system's life to some extent.

2.2.2.1.2 Overhaul

Overhaul activities are very important and necessary tasks in repairable systems' TTL. Overhauling includes disassembling the whole system, inspection of each component, and again reassembling the system. It is assumed that overhauling rejuvenates the system to such an extent that it brings the system to as good as new condition. Since overhauling is not performed after any failure, it comes under the PM category. It is also called "capital maintenance" in some industries.

2.2.2.2 Predictive Maintenance

In predictive maintenance (Ebeling, 2004), we predict the failures before it occurs by continuously monitoring the system using different sensors. Sensors continuously monitor system health and provide indications before failures occur. It is also called condition-based maintenance (CBM). However, CBM is mainly recommended for critical systems, since it is not cost-justifiable or feasible on all systems. However, in this book, we have only considered preventive and corrective maintenance for further analysis.

2.3 Classifications of Maintenance Categories

Having described this classification in brief in the first chapter, here, we now discuss the parametric methods for reliability estimation of performance parameters in case of perfect, minimal, and imperfect maintenance.

Reliability modeling of repairable systems is done with the help of various parametric and nonparametric approaches. In this work, we have mainly focused on parametric approaches considering CM and PM. These two major maintenance categories can be further classified based on the degree by which it restores the system to its operating condition (Rai and Bolia, 2014; Pham and Wang, 1996; Rajiv N. Rai and Sharma, 2017), namely, perfect maintenance, minimal maintenance, and imperfect maintenance. Let us discuss each maintenance category along with the parametric approaches to deal with them. There are various parametric approaches available in the literature for each maintenance category, but we have mainly discussed the homogeneous Poisson process (HPP), non-homogeneous Poisson process (NHPP), and generalized renewal process (GRP) in the subsequent sections.

2.3.1 Perfect Maintenance

When the system is brought to "as good as new" (AGAN) condition after performing any maintenance action, it is called perfect maintenance. On failure or during maintenance, replacing the system with a new one is categorized as perfect CM and perfect PM, respectively. Other than SPM, repairable systems undergo a PM activity called overhauling which may bring it to AGAN condition (it is an assumption). A detailed discussion about overhauling is covered in Chapter 3.

2.3.1.1 Parametric Approach for Perfect Maintenance

Renewal process (RP) (Briš and Byczanski, 2017) is generally applied for reliability analysis of repairable systems which undergo perfect maintenance, provided that the time to failures (TTFs) of systems are independently and identically distributed. A special case of RP called the homogeneous Poisson process (HPP), where distribution is exponential (Muralidharan, 2008) and the intensity function remains constant, is used for the analysis purpose of such systems. For HPP, the expected number of failures $E[N(t)]$ in time interval $[0, t]$ is given by λt. Thus, the intensity function in this case becomes:

$$u(t) = \frac{d}{dt}E[N(t)] = \lambda \tag{2.15}$$

The constant intensity function of HPP restricts it to apply when there is an improving or deteriorating trend in the failure profile of repairable systems. To capture such trends, a nonconstant intensity function is required to be used.

2.3.2 Minimal Maintenance

When the performed maintenance action brings the system to "as bad as old condition" (ABAO) or at the same stage where it was before the maintenance, it is known as minimal maintenance action. It is a more practical assumption than AGAN to deal with repairable systems.

2.3.2.1 A Parametric Approach to Deal with Minimal Maintenance

As discussed earlier, the TBFs of repairable systems do not follow the i.i.d. assumption since they can fail more than once followed by a repair action. To model such conditions, a nonhomogeneous Poisson process (NHPP) (Crow, 1975; Finkelstein and Shafiee, 2017) is used wherein the intensity function is nonconstant. The intensity function of NHPP can capture the improving and deteriorating trend of the repairable system's failure profile as the expected number of failures in a small interval [0, t] is given by

$$E[N(t)] = \int_0^t u(t)dt \tag{2.16}$$

Here $u(t)$ is approximated by the power law model (Crow, 1975), which is given as follows:

$$u(t) = a \times \beta \times t^{\beta-1}; a > 0, \beta > 0 \tag{2.17}$$

where a and β are the scale and shape parameters of the model, respectively.

Equation (2.17) is used to estimate the intensity function at each successive failure time that occurs in a particular system. However, the intensity function of the very first failure time is approximated by Weibull distribution as shown in Eq. (2.18).

$$u(t) = \frac{\beta}{\eta}\left(\frac{t}{\eta}\right)^{\beta-1} \tag{2.18}$$

Comparing Eq. (2.17) with Eq. (2.18),

$$a = \frac{1}{\eta^\beta};$$

hence

$$E[N(t)] = \int_0^t u(t)dt = at^\beta \text{ or } \left(\frac{t}{\eta}\right)^\beta \tag{2.19}$$

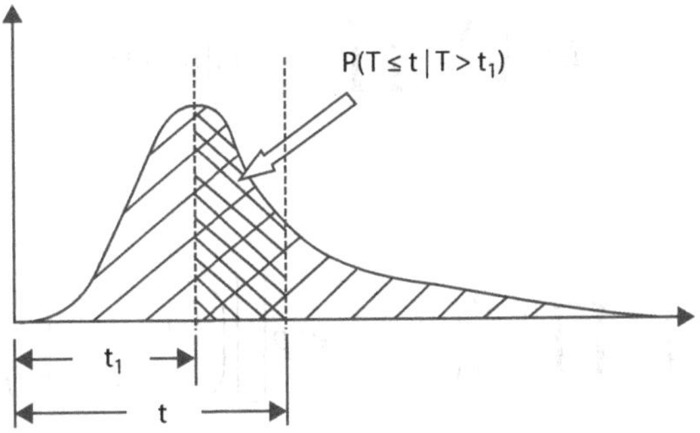

Figure 2.4 Conditional probability density function for NHPP (Yanez, Joglar, and Modarres, 2002).

To obtain the model parameters, the following illustration of conditional PDF is used as shown in Figure 2.4.

From Figure 2.4:

$$P(T \leq t | T > t_1) = \frac{F(t) - F(t_1)}{R(t_1)} = 1 - \frac{R(t)}{R(t_1)} \quad (2.20)$$

Where F (·) and R (·) are the probability of component failure and the reliability at the respective times. Thus, for the ith failure in the lth system ($l = 1,2, \ldots K$),

$$F(t_{li}) = 1 - exp\left[\left(\frac{t_{l,i-1}}{\eta}\right)^{\beta} - a\left(\frac{t_{li}}{\eta}\right)^{\beta}\right] \quad (2.21)$$

The PDF of the ith failure given that $(i-1)^{th}$ failure occurred at time t_{i-1} will be:

$$f\left((t_{li} \mid t_{l,i-1})\right) = \frac{\beta}{\eta}\left(\frac{t_{li}}{\eta}\right)^{\beta-1} exp\left[\left\{\left(\frac{t_{l,i-1}}{\eta}\right)^{\beta} - \left(\frac{t_{li}}{\eta}\right)^{\beta}\right\}\right] \quad (2.22)$$

The NHPP model parameters can be obtained with the help of the maximum likelihood estimation (MLE) method. The general form of the likelihood function is given by:

$$L = \prod_{l=1}^{K} \prod_{i=1}^{p_l} f((t_{li}|t_{l,i-1})) \tag{2.23}$$

Thus,

$$L = \prod_{l=1}^{K} \prod_{i=1}^{p_l} \frac{\beta}{\eta}\left(\frac{t_{li}}{\eta}\right)^{\beta-1} exp\left[\left\{\left(\frac{t_{l,i-1}}{\eta}\right)^{\beta} - \left(\frac{t_{li}}{\eta}\right)^{\beta}\right\}\right] \tag{2.24}$$

p_l is the number of failures in the lth system.

Taking log on both sides:

$$\begin{aligned} logL = log(\beta)\sum_{l=1}^{K} p_l + (\beta-1)\sum_{l=1}^{K}\sum_{i=1}^{p_l} logt_{li} - \beta log(\eta)\sum_{l=1}^{K} p_l \\ + \frac{1}{\eta^{\beta}}\sum_{l=1}^{K}\sum_{i=1}^{p_l}\left(t_{l,i-1}{}^{\beta} - t_{l,i}{}^{\beta}\right) \end{aligned} \tag{2.25}$$

Differentiating of Eq. (2.25) w.r.t. the model parameters (η and β in this case) and equating it to zero provides the MLEs for the parameters. The MLEs are provided in Appendix A. The parameters also can be obtained by maximizing the log-likelihood equation.

Thus, from the above discussion, it can be concluded that the assumption of interarrival TBFs follows a conditional Weibull probability distribution which means that the arrival of the ith failure is conditioned on the accumulative operating time up to the $(i - 1)^{th}$ failure. This conditionality also reveals the fact that the system is restored to ABAO condition after the $(i - 1)^{th}$ repair, and hence, the repair action has not restored the life of the system to any extent.

2.3.3 Imperfect Maintenance

The ideology behind imperfect maintenance is that the performed maintenance actions can bring the system to some intermediate stage, i.e., between AGAN and ABAO conditions. The possibilities of maintenance being perfect and minimal could be the special cases of imperfect maintenance. The state, where the system is brought to, depends on the quality of maintenance performed on the system. In the literature, various arithmetic reductions of age (ARA) and arithmetic reduction in intensity (ARI)-based models (de Toledo *et al.*, 2015) are available to deal with imperfect maintenance. The arithmetic reduction of age models (Doyen and Gaudoin, 2004) assume that system's failure intensity at actual age (t) is equal to its intensity function at virtual age (V_t), where $V_t \leq t$. Similarly, between two consecutive failures, its failure intensity is horizontally parallel to its initial intensity. Hence, in ARA models, it is assumed that the repair action reduces the increment in system age. Furthermore, ARI models (Doyen and Gaudoin, 2004) assume that repair actions reduce not only the virtual age but also the failure intensity function of the system. Hence, ARI models assume that, between two consecutive failures, its failure intensity is vertically parallel to its initial intensity and repair reduces the increment in failure intensity.

In this book, we mainly concentrate on the ARA-based virtual age models which are proposed by Kijima and Sumita (1986) considering imperfect CM and Nasr, Gasmi, and Sayadi (2013) considering imperfect SPM and CM.

2.3.3.1 *Parametric Approaches to Deal with Imperfect Maintenance*

The generalized renewal process (GRP) (das Chagas Moura *et al.*, 2014) is generally used for reliability analysis of repairable systems when imperfect maintenance is considered. The generalized renewal process was introduced by Kijima and Sumita (1986) and Kijima (1989) in which they proposed two virtual age-based models, namely, Kijima I (KI) and Kijima II (KII), to identify the intermediate state of the system after CM is performed. The model introduces a new factor called the repair effectiveness index (REI) "q" in the power law process which represents the quality of repair. The KI and KII models are explained below:

- *Kijima I model (KI):* According to the KI model, the repair is assumed to restore the accumulated damage during the time between the ith and $(i-1)^{th}$ failure. The virtual age formula for the KI model is expressed in Eq. (2.26).

$$V_{li} = V_{l,i-1} + qy_{li};\ (i = 1,2, \ldots.\ p)(l = 1,2, \ldots\ K) \tag{2.26}$$

  where y_{li} is the time between failures of the lth system and q is the repair effectiveness index (REI).
- *Kijima II model (KII):* According to the KII model, the ith repair action at any point of time restores the entire accumulated age since new. The virtual age formula for the KII model is expressed in Eq. (2.27).

$$V_{li} = q(V_{l,i-1} + y_{li}); \; (i = 1,2, \ldots. p) \; (l = 1,2, \ldots K) \tag{2.27}$$

$$q = \begin{cases} 0 \;; & \text{for perfect repair (as good as new condition)} \\ 0 < q < 1; & \text{for imperfect repair (intermediate state)} \\ 1 \;; & \text{for minimal repair (as bad as old condition)} \end{cases} \tag{2.28}$$

The assumptions of the above models are as follows: (1) the first failure of the system follows the Weibull distribution, and (2) repair time is negligible. Depending on the value of q, GRP can be converted into other models such as HPP and NHPP as shown in Figure 2.5.

The CDF, conditional probability density function (Yu, Song, and Cassady, 2008), likelihood function, and reliability function considering GRP are given as follows:

CDF:

$$F(t_{li}) = 1 - exp\left[\left(\frac{V_{l,i-1}}{\eta}\right)^{\beta} - \left(\frac{V_{l,i-1} + y_{li}}{\eta}\right)^{\beta}\right]; \tag{2.29}$$

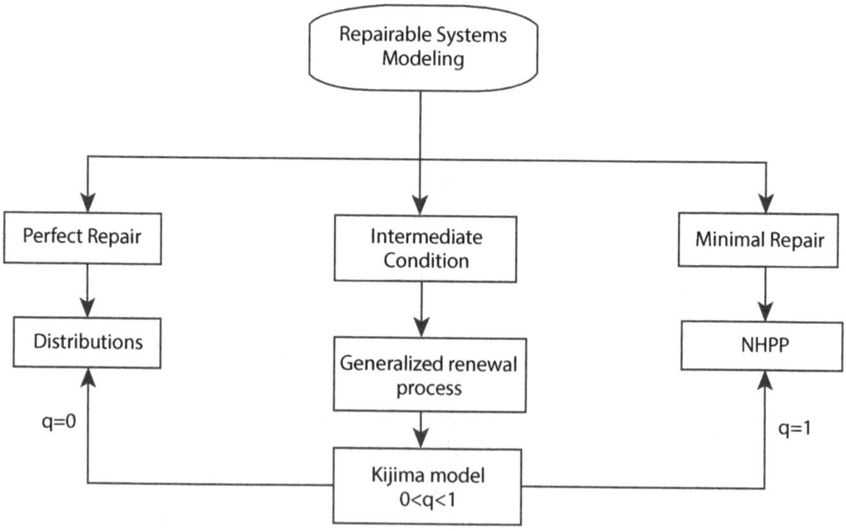

Figure 2.5 Conversion of GRP into other models.

Conditional probability density function:

$$f(t_{li}|t_{l,i-1}) = \frac{\beta}{\eta}\left(\frac{V_{l,i-1}+y_{li}}{\eta}\right)^{\beta-1} exp\left[\left(\frac{V_{l,i-1}}{\eta}\right)^{\beta} - \left(\frac{V_{l,i-1}+y_{li}}{\eta}\right)^{\beta}\right]; \tag{2.30}$$

Here, in Eq. (2.30), V_{li} is given by either Eq. (2.26) or Eq. (2.27) for the KI and KII models, respectively.

Likelihood function:

$$L(Data|Parameters)$$

$$= \prod_{l=1}^{K}\prod_{i=1}^{p_l} \frac{\beta}{\eta}\left(\frac{V_{l,i-1}+y_{li}}{\eta}\right)^{\beta-1} exp\left[\left(\frac{V_{l,i-1}}{\eta}\right)^{\beta} - \left(\frac{V_{l,i-1}+y_{li}}{\eta}\right)^{\beta}\right] \tag{2.31}$$

Taking log on both sides

$$\begin{aligned} log\ L = {} & log(\beta)\textstyle\sum_{l=1}^{K} p_l + (\beta-1)\sum_{l=1}^{K}[\sum_{i=1}^{p_l} log(V_{l,i-1}+x_{l,i})] \\ & - \beta\ log(\eta)\textstyle\sum_{l=1}^{K} p_l + \sum_{l=1}^{K}\left[\sum_{i=1}^{p_l}\left[\left(\frac{V_{l,i-1}}{\eta}\right)^{\beta} - \left(\frac{V_{l,i-1}+x_{l,i}}{\eta}\right)^{\beta}\right]\right] \end{aligned} \tag{2.32}$$

Where K is number of systems if multiple repairable systems are considered and T test terminated time. Differentiating of Eq. (2.32) w.r.t. the model parameters (η, β, and q in this case) and equating it to zero provides the MLEs for the parameters. The equations for MLEs for the KI and KII models are provided in Appendix A.

Reliability function for the first failure:

$$R(t_{l1}) = exp\left[-\left(\frac{y_{l1}}{\eta}\right)^{\beta}\right]; \tag{2.33}$$

Reliability function from the second failure onward:

$$\mathrm{R}(t_{li}) = exp\left[\left(\frac{V_{l,i-1}}{\eta}\right)^{\beta} - \left(\frac{V_{l,i-1} + y_{li}}{\eta}\right)^{\beta}\right]; \tag{2.34}$$

Essentially, GRP was used by the researchers to model imperfect CM (Tanwar, Rai, and Bolia, 2014), and accordingly, likelihood function and maximum likelihood estimators (MLEs) for the KI and KII models were proposed (Yanez, Joglar, and Modarres, 2002; Mettas and Zhao, 2005) to estimate the model parameters, i.e., a, β, and q. Later, Nasr, Gasmi, and Sayadi (2013) introduced the concept of imperfect PM and CM in the Kijima models. Nasr, Gasmi, and Sayadi (2013) introduced the modified virtual age equation for the KI and KII models as given in Eqs. (2.35) and (2.36), respectively, using two quality indices separately for CM and SPM, namely, q_{CM} and q_{SPM}, and also proposed the likelihood function and MLEs for the model parameters as shown in Eq. (2.37). Here, the PM considered by the authors was normal scheduled preventive maintenance (SPM) which includes small timely inspections as explained earlier.

$$V_{li} = \sum_{l=1}^{K} \sum_{j=1}^{i} q_{CM}^{\delta_j} q_{SPM}^{1-\delta_j} x_{lj} \tag{2.35}$$

$$q_{CM} = \begin{cases} 0\ ; & \textit{for perfect CM(as good as new condition)} \\ 0 < q_{CM} < 1; & \textit{for imperfect CM(intermediate state)} \\ 1\ ; & \textit{for minimal CM(as bad as old condition)} \end{cases}$$

$$q_{SPM} = \begin{cases} 0\ ; & \textit{for perfect SPM (as good as new condition)} \\ 0 < q_{SPM} < 1; & \textit{for imperfect SPM (intermediate state)} \\ 1\ ; & \textit{for minimal SPM (as bad as old condition)} \end{cases}$$

$x_{li} = t_{li} - t_{l,i-1}$, i.e., the time between interventions (either SPM or CM) and

$$\delta_{li} = \begin{Bmatrix} 1 \ \ if\ CM \\ 0\ if\ SPM \end{Bmatrix}$$

$$V_{li} = \sum_{j=1}^{i} q_{CM}^{(i-j+1)\delta_{lj}} q_{SPM}^{(i-j+1)1-\delta_{lj}} x_{lj} \tag{2.36}$$

$$L\,(x_1, x_2, \ldots\ldots x_n | \beta, \eta, q_{CM}, q_{SPM}) = \sum_{l=1}^{K} \prod_{i=1}^{n_l} \left[\left\{ \frac{\beta}{\eta} \left(\frac{V_{l,i-1} + x_{li}}{\eta} \right)^{\beta-1} \right\}^{\delta_{li}} \right] \prod_{i=1}^{n_l} exp \left\{ \left(\frac{V_{l,i-1}}{\eta} \right)^{\beta} - \left(\frac{V_{l,i-1} + x_{li}}{\eta} \right)^{\beta} \right\} \tag{2.37}$$

Taking log on both sides:

$$log\ L = log(\beta) \sum_{l=1}^{K} [\sum_{i=1}^{n_l} (\delta_{l,i})] + (\beta - 1) \sum_{l=1}^{K} [\sum_{i=1}^{n_l} (\delta_{l,i})\, log(V_{l,i-1} + x_{l,i})] - \beta\ log(\eta) \sum_{l=1}^{K} [\sum_{i=1}^{n_l} (\delta_{l,i})] + \sum_{i=1}^{K} \left[\sum_{i=1}^{n_l} \left[\left(\frac{V_{l,i-1}}{\eta} \right)^{\beta} - \left(\frac{V_{l,i-1} + x_{l,i}}{\eta} \right)^{\beta} \right] \right] \tag{2.38}$$

Here, in Eqs. (2.37) and (2.38), V_{li} is given by either Eq. (2.35) or Eq. (2.36) for the KI and KII models, respectively, and n_l is the number of interventions in the lth system (due to SPM or CM). Differentiating the log of Eq. (2.38) w.r.t. the model parameters (η, β, q_{CM}, and q_{SPM} in this case) and equating it to zero provides the MLEs for the parameters. The equations for MLEs for the KI and KII models proposed by Nasr, Gasmi, and Sayadi (2013) are given in Appendix B.

Modeling of failure times for repairable systems can be done using various models as available in the literature (Rai, Chaturvedi *et al.*), but analyzing failure modes (FMs) and their effects under different conditions, especially when the collected data are censored, is an uphill task. To analyze the failure data concerning one particular FM, other FMs are considered as censored. In this book, we have also focused on the techniques for analyzing censored data of repairable systems. The next section is dedicated to the basic concept of data censoring.

2.4 Concept of Censoring

Analysis of censored data is one of the very important and unavoidable events while analyzing repairable system failure data. To correctly estimate the statistical model parameters, complete information about system failures is required. The industries generate the failure data either through reliability testing or directly from the field. The most common problem that arises with generated data is "censoring." Collected data get censored when units are removed from consideration before failure or the test is completed prior to all units failing (Ebeling, 2004). Censoring is a property of data collection (Nelson, 2003) and can be broadly classified into two categories: (1) singly censored and (2) multiply censored (Ebeling, 2004). Singly censored data can be again categorized into left or right censoring. If the interested event is the failure of a system, left censoring occurs when failures take place before the start of the test. In contrast, right censoring takes place when failures take place after the completion of the test. Hence, in both cases, we do not get complete information about the failures. In right censoring, the test termination point is based on a fixed length of time or fixed number of failures and is called type I censoring (time truncated) or type II censoring (failure truncated) accordingly (Dwivedi and Sachdeva, 2016; Dai and Wang, 2016). Multiply censoring occurs when running and failure times get mixed up. Such situations occur when (1) test time or operating time differs among the censored units (removed but operating). Censored units are removed at various times from the sample or have gone into service at different times; (2) failures occur because of FMs which are not of interest (Ebeling, 2004; Nelson, 2009). Thus, censored data always produce some ambiguity in the reliability analysis of both nonrepairable and repairable systems (Rigdon and Basu, 2000).

The analysis of censored data for repairable systems follows a different approach altogether as compared with nonrepairable systems since the repairable systems are subjected to various repair (CM) actions (H. Ascher and Feingold, 1984) after failure along with the SPM actions at fixed intervals. Hence, modeling of censored data for repairable systems is essential to get realistic performance parameters. A detailed discussion on data censoring techniques for repairable systems is provided in Chapter 4.

Having discussed the available approaches/models related to modeling and analysis of repairable systems (Sharma, Sahu, and Rai, 2022; Sharma and Rai, 2018b), the next section discusses the limitations of present approaches in MRO to deal with reliability modeling and analysis of repairable systems.

2.5 Problems Faced by the Industries: Present Scenario

Repairable systems are subjected to periodic overhauls that involve several complex and major activities during the life cycle of the equipment. In most industries, MRO facilities assume that the overhaul activities restore the system to AGAN condition which is an unrealistic assumption in the case of repairable systems and can be considered as one of the major limitations. The available approaches of imperfect maintenance modeling consider CM and PM as imperfect. However, PM consists of SPM and overhaul, and both activities are quite different as discussed previously. Considering merely CM and PM, without differentiating the activities involved in it, may mislead the organization during the reliability analysis of the system. In addition, it becomes imperative to consider all the interventions in the form of SPM, overhaul, and CM as imperfect simultaneously to portray the complete maintenance activities with precision.

Another problem often confronted by industries is to estimate and analyze the performance parameters (Ebeling, 2004) from failure data when imperfect maintenance and FM-wise censored data are required to be considered to obtain the parameters for each FM separately. Such analysis could be very helpful for MRO facilities in conducting FMEA of the systems.

Maintenance, repair, and overhaul facilities also face problems in identifying high failure rate components (HFRCs) that are subjected to defect investigations and further reliability analysis. The selection of HFRCs is done mainly through experts' opinions. As a result, most of the time, the MRO gets overloaded with systems that do not come under the HFRC category. On the other hand, if the deserving component is not declared as HFRC, it may lead to unsuccessful missions or severe causalities. A scientific methodology is required to be developed for designating the components as HFRCs.

Furthermore, the present traditional/classic maintenance policies, having fixed maintenance interval times, may not be that effective in the case of large and complex repairable systems since their deterioration increases with age. This may cause the system to get identified as HFRC earlier. Hence, another solution, to avoid unnecessary HFRC burden on MRO units, could be developing a maintenance policy that can update the SPM and overhaul intervals according to the failure profile of the systems. Such maintenance policy may prevent the system to get deteriorated fast and also prevent them to fall under the HFRC category. Presently, most of the MRO facilities in the industries follow the classic maintenance policies for repairable systems.

Furthermore, the MRO facilities are also responsible for improving the quality of maintenance done on repairable systems. The effectiveness of these maintenance actions could be dependent on numerous subjective factors and requires qualitative analysis for a better appreciation of repair effects and performance. A quantitative assessment of repair quality is necessary but not sufficient to analyze it completely. Hence, a qualitative analysis of maintenance activities, which can improve the quality of repair actions and provide remedial solutions, is highly demanded in the industries at the present time. Such study could help in appreciating the strengths as well as the weaknesses present in the organization in this regard.

Hence, by addressing the above-discussed limitations, MRO facilities can be further equipped with more scientific models and methodologies to deal with repairable systems reliability analysis. The primary purpose of this book is to provide the solution of some practical issues confronted by various industries while dealing with repairable systems. The subsequent sections highlight the observations on dealing with maintenance issues and analyzing the lifetime data for repairable systems in detail.

2.5.1 Observation 1

The available literature is more inclined toward fixing the overhaul intervals with cost optimization considering imperfect maintenance which generally includes SPM or CM as imperfect. However, with the optimal number of overhauls, the quality of the overhaul is also equally important to achieve the desired results in terms of better performance of the system. The various virtual age models and maintenance policies developed in the literature consider imperfect PM which could be either SPM or overhaul. However, the activities done in both types of maintenance are very different as discussed earlier. Few research works assume overhaul as imperfect considering fixed improvement factor with minimal CM. On the other hand, some consider only SPM and CM as imperfect. However, during the life cycle of a repairable system, each maintenance activity (i.e., CM, SPM, and overhaul) can be imperfect at the same time. The Kijima models as well only consider

CM as imperfect. Hence, the assumption of a perfect overhaul in MRO facilities may lead to an overestimation of the performance parameters of repairable systems. Thus, this assumption is required to be waived off for more realistic results while estimating various performance parameters of repairable systems. Chapter 3 is dedicated toward offering a methodology that can model all the interventions deeming them imperfect.

2.5.2 Observation 2

Nowadays, with the help of advanced technology, the industries keep records of each failure along with their respective FMs, but still, they face difficulties while analyzing the data for a particular FM keeping others censored. The available Kijima (GRP)-based parametric methods use all failure times irrespective of their FMs without considering the interventions due to SPM. More precise failure analysis can be performed by identifying and segregating various FMs from the failure data. The repairable system data censoring analysis acquires more complicacy when parameter estimation is required to be done based on the FM approach in conjunction with imperfect SPM interventions. The literature is lacking in providing the methodology to deal with such cases. The book contributes an FM-based censored data analysis method in Chapter 4.

2.5.3 Observation 3

In the industries, most of the large and complex systems are treated as repairable systems and, hence, need MRO facilities. The facilities of MRO play a major role in taking obligatory actions on the components subjected to frequent failures. Arrivals of the same component due to repeated failures and other interventions, as acknowledged through the defect reporting and investigation database, are recognized and designated as high failure rate components (HFRCs) (Rai and Bolia, 2014). To identify HFRCs and reduce the risk and cost involved in their maintenance, they require timely observations of their degradation profile. In such industries, the identification of HFRCs is generally based on expert's intuitions and experience which could be a reason for designating undesirable components as HFRC. Incorrectly designated HFRCs impose excessive or undue load on production lines, and there could be a predicament of lost opportunities. Thus, by acknowledging the limitations of the existing methods, a methodology for identifying HFRCs is necessary and also demanded in industries. This book endeavors to provide a methodology for identifying HFRCs as explained in detail in Chapter 5.

2.5.4 Observation 4

Maintenance policies are essential in achieving the systems' operational effectiveness at minimum cost and downtime. The available literature provides various maintenance policies for repairable systems concentrating on cost and downtime optimization. However, in the case of repairable systems, maintenance plans could be more effective if they are based on the failure profile of the equipment to cater for its wear out rather than following the traditional methods. The situation can become even worse if it is planned to give a life extension to the equipment based on the stipulated life. Since the traditional/classic maintenance policies are not based on mean residual life (MRL) or risk of failures of the equipment, the life extension provided is not realistic and the equipment faces serious repercussions in terms of major failures, some of them being catastrophic. Hence, such critical and complex repairable systems require such advanced maintenance policy which plans the next SPM and overhaul activity time based on the failure profile of the systems so that the risk of failure is not increased. This could also be a solution to prevent the systems from being HFRC which can again help in strengthening MRO facilities in industries. The available literature is lacking in providing such maintenance policy. Chapters 6 and 7 provide progressive- and age-based maintenance policies for repairable systems, respectively.

2.5.5 Observation 5

A system may fail due to various failure modes requiring different types of repair actions after each failure. Each repair action is a function of various subjective factors affecting that particular repair. Hence, merely, a number may not always be a true representative of repair quality. Thus, differentiating repair actions purely based on numerical values may not provide a holistic view of the complete repair process of a system. Quantitative assessment of REI is necessary but not sufficient to provide indications to incorporate remedial measures for improving the performance of a system. Thus, an organization must understand the meaning of REI both qualitatively and quantitatively to portray a complete picture of the repair quality. In other words, REI could be dependent on many subjective factors, and the dependability of REI on such factors is required to be modeled for a better understanding of repair effects and performance. In the literature, adequate attention is given to the quantitative analysis of "q" and estimation of model parameters (η, β, and q) with the help of the maximum likelihood function, but it lacks in providing methodology which can be utilized for the qualitative assessment of REI. Chapter 8 is dedicated toward offering methodologies for the qualitative assessment of REI.

2.6 Chapter Summary

The basic understanding of MRO was provided in the first chapter. This chapter further supplements the comprehension of terminologies used in reliability modeling of repairable systems along with their mathematical expressions for a better appreciation of the advanced techniques of repairable systems reliability analysis with their applications to the imperfect maintenance as covered in the subsequent chapters. To understand various activities such as CM, SPM, and overhaul, a brief explanation is provided with examples. In addition, the various categories of maintenance like perfect, minimal, and imperfect maintenance are described to build up the basic understanding of repairable system maintenance modeling. Lastly, the chapter focuses on real-life industrial problems for which the solutions are provided in the subsequent chapters.

References

Ascher, H. and Feingold, H., *Repairable systems reliability*, Hoboken, New Jersey, U.S., 1984, https://books.google.co.in/books?id=mYZPAQAAIAAJ.

Ascher, H.E., Repairable systems reliability, in: *Encyclopedia of Statistics in Quality and Reliability*, vol. 4, 2008.

Briš, R. and Byczanski, P., On innovative stochastic renewal process models for exact unavailability quantification of highly reliable systems. *Proc. Inst. Mech. Eng. Part O: J. Risk Reliab.*, 231, 6, 617–27, 2017.

Chagas Moura, M.D., Droguett, E.L., Firmino, P.R.A., Ferreira, R.J., A competing risk model for dependent and imperfect condition–based preventive and corrective maintenances. *Proc. Inst. Mech. Eng. Part O: J. Risk Reliab.*, 228, 6, 590–605, 2014.

Crow, L.H., Reliability analysis for complex, repairable systems, 1975, ARMY MATERIEL SYSTEMS ANALYSIS ACTIVITYABERDEEN PROVING GROUND MD, *Reliability and Biometry*, 13, 6, 379–410, 1975.

Dai, H. and Wang, H., *Analysis for Time-to-Event Data under censoring and truncation*, Academic Press, 2016.

Doyen, L. and Gaudoin, O., Classes of imperfect repair models based on reduction of failure intensity or virtual age. *Reliab. Eng. Syst. Saf.*, 84, 1, 45–56, 2004.

Dwivedi, N. and Sachdeva, S., Survival analysis: A brief note. *J. Curr. Res. Sci. Med.*, 2, 2, 73, 2016.

Ebeling, C.E., *An introduction to reliability and maintainability engineering*, Tata McGraw-Hill Education, New York, United States, 2004.

Finkelstein, M. and Shafiee, M., Preventive maintenance for systems with repairable minor failures. *Proc. Inst. Mech. Eng. Part O: J. Risk Reliab.*, 231, 2, 101–8, 2017.

Jardine, A.K.S. and Tsang, A.H.C., *Maintenance, replacement, and reliability: Theory and applications*, CRC press, Boca Raton, Florida, 2005.

Kijima, M., Some results for repairable systems with general repair. *J. Appl. Probab.*, 26, 1, 89–102, 1989.

Kijima, M. and Sumita, U., A useful generalization of renewal theory: Counting processes governed by non-negative markovian increments. *J. Appl. Probab.*, 23, 1, 71–88, 1986.

Lindqvist, B.H., On the statistical modeling and analysis of repairable systems. *Stat. Sci.*, 21, 4, 532–51, 20, 2006.

Mettas, A. and Zhao, W., Modeling and analysis of repairable systems with general repair, in: *Annual Reliability and Maintainability Symposium, 2005. Proceedings*, pp. 176–82, IEEE, Alexandria, VA, USA, 2005.

Muralidharan, K., A review of repairable systems and point process models. *ProbStat Forum*, 1, 26–49, 2008.

Nasr, A., Gasmi, S., Sayadi, M., Estimation of the parameters for a complex repairable system with preventive and corrective maintenance. *International Conference On Electrical Engineering and Software Applications (ICEESA)*, pp. 1–6, 2013.

Nelson, W.B., *Recurrent events data analysis for product repairs, disease recurrences, and other applications*, vol. 10, SIAM, Philadelphia, PA, 2003.

Nelson, W.B., *Accelerated testing: Statistical models, test plans, and data analysis*, vol. 344, John Wiley & Sons, Hoboken, New Jersey, U.S., 2009.

Pham, H. and Wang, H., Imperfect maintenance. *Eur. J. Oper. Res.*, 94, 3, 425–38, 1996.

Rai, R.N. and Bolia, N., Availability based optimal maintenance policies in military aviation. *Int. J. Perform. Eng.*, 10, 6, 641–48, 2014.

Rai, R.N., Chaturvedi, S.K., Bolia, N., *Repairable systems reliability analysis: A comprehensive framework*, John Wiley & Sons, Hoboken, New Jersey, U.S., 2020.

Reliawiki, *Introduction to repairable systems*, Hoboken, New Jersey, U.S., 2015, http://reliawiki.com/index.php/Introduction_to_Repairable_Systems.

Rigdon, S.E. and Basu, A.P., *Statistical methods for the reliability of repairable systems*, Wiley New York, Hoboken, New Jersey, U.S.,2000.

Rolnitzky, A., *Reliability of repairable systems vs. non-repairable systems*, Las Vegas, Nevada, 2011, https://opsalacarte.com/reliability-of-repairable-systems-vs-non-repairable-systems/.

Saaty, T.L., Decision making with the analytic hierarchy process. *Int. J. Serv. Sci.*, 1, 1, 83–98, 2008.

Sharma, G., Sahu, P.K., Rai, R.N., Imperfect maintenance and proportional hazard models: A literature survey from 1965 to 2020. *Life Cycle Reliab. Saf. Eng.*, 11, 1, 87–103, 2022.

Rai, R.N. and Sharma, G., Goodness-of-fit test for generalised renewal process. *Int. J. Reliab. Saf.*, 11, 1–2, 116–31, 2017.

Sharma, G. and Rai, R.N., Reliability modeling and analysis of environmental control and life support systems of space stations: A literature survey. *Acta Astronaut.*, 155, 238–46, 2018b.

Tanwar, M., Rai, R.N., Bolia, N., Imperfect repair modeling using kijima type generalized renewal process. *Reliab. Eng. Syst. Saf.*, 124, 24–31, 2014.

De Toledo, M.L.G., Freitas, M.A., Colosimo, E.A., Gilardoni, G.L., ARA and ARI imperfect repair models: Estimation, goodness-of-fit and reliability prediction. *Reliab. Eng. Sys. Saf.*, 140, 107–15, 2015.

Yahyatabar, A. and Najafi, A.A., Condition based maintenance policy for series-parallel systems through proportional hazards model: A multi-stage stochastic programming approach. *Comput. Ind. Eng.*, 126, 30–46, 2018.

Yanez, M., Joglar, F., Modarres, M., Generalized renewal process for analysis of repairable systems with limited failure experience. *Reliab. Eng. Syst. Saf.*, 77, 2, 167–80, 2002.

Yu, P., Song, J.J., Cassady, C.R., Parameter estimation for a repairable system under imperfect maintenance, in: *Annual Reliability and Maintainability Symposium, 2008. RAMS 2008*, pp. 428–33, IEEE, Las Vegas, Nevada, 2008.

3

Imperfect Overhaul Virtual Age Model

3.1 Introduction

Maintenance scheduling of repairable systems is considered to be an essential task because the cost of production loss due to failure is significant, and the cost of owning such repairable systems is generally lesser when they are properly taken care of during their useful life. The quality of maintenance actions affects the performance of the system to a large extent. As discussed in Section 2.2, maintenance actions can be majorly divided into two categories, namely, preventive (under the proactive maintenance category; predictive maintenance is not in the scope of this book) and corrective maintenance (PM and CM). Preventive maintenance is again categorized into two segments based on the activities done on the system, i.e., SPM and overhaul. One can say that PM is a total of all those activities which aim to maintain, improve, and prevent the system from future failures. Let us discuss each of them one by one.

3.1.1 Activities during SPM

Scheduled preventive maintenance essentially includes lubrication, calibrations, minor adjustments, oil filling or changing, filter cleaning, replacement of wearing parts like bearings, and running checks and tests on subsystems and components. The periodicity or frequency for these SPMs is generally recommended by the manufacturer, for example, daily, weekly, or every 50 h.

3.1.2 Activities during Overhaul

Overhauling involves the dismantling of the system, functional testing of components and subsystems, calibration and readjustment, replacement of damaged parts, servicing activities (such as cleaning of components;

Garima Sharma and Rajiv Nandan Rai. Advanced Techniques for Maintenance Modeling and Reliability Analysis of Repairable Systems, (65–80) © 2023 Scrivener Publishing LLC

greasing and replenishment of consumables, such as fuel, lubricating oil, and compressed air), and finally reassembling of the system.

The periodicity of overhaul is primarily decided by the manufacturer, but it could be reviewed later when maintenance cost or hazard rate is found to increase significantly. In many industries, it gets preponed for the sake of convenience, when some major failure and the elapsed time from the preceding overhaul of the system are close enough to the statutory requirement.

Most of the industries (MRO facilities) and also the literature consider overhaul as the perfect PM action which brings the system to the AGAN condition. However, some consider it as restorative action but with some assumptions as deliberated in Section 2.5.1. Thus, in this chapter, we have concentrated on developing an ARA-based virtual age model which would be able to consider CM, SPM, and overhaul (each intervention done on the system) as imperfect simultaneously. Let us first understand why there is a need of considering overhaul activity as imperfect.

3.2 Need for an Imperfect Overhaul

An overhaul is considered to be an imperative PM activity for complex and critical systems. Hence, the analysis of overhaul quality becomes important. Most of the available literature and industry manuals assume that overhauling rejuvenates the system completely and brings it to AGAN condition. But, in reality, considering overhaul, which is a very important and massive activity done on repairable systems, as perfect is practically not possible. Overhaul improves the condition of the system but does not bring the system to its original (new) condition (Seo and Bai, 2004; Pascual and Ortega, 2006). To support this statement, let us consider the following case study of two jet aero engine components, namely, turbo starter and plunger pump. Turbo starters are used to start the main aero engines, whereas plunger pumps are reciprocating positive displacement pumps designed to supply fuel to the combustion chambers and jet pipes. Both components are subjected to SPM of 100 and 200 h with a different overhaul periodicity. The turbo starters and plunger pumps have undergone three and two overhaul cycles, respectively. Failure times, time to overhaul, and number of failures of both components during their respective overhaul cycles are shown in Table 3.1.

Before discussing the methodology, let us now understand the need for an imperfect overhaul. Let us now observe the number of failures in each

Table 3.1 Data description of the turbo starter and plunger pump.

Component name	Failure times	Overhaul time	No. of failures in each overhaul cycle		
			First	Second	Third
Turbo starter	259, 464, 578, 744, 841, 884, 953, 1262, 1337, 1347, 1386, 1423, 1487, 1567, 1635, 1689, 1762, 1845, 1862, 1956, 1983, 2073, 2190, 2276, 2366, 2399, 2436, 2443, 2489, 2573, 2590, 2615, 2694, 2846, 2874, 2888, 2930, 2971, 2990	1000 h (SPM 100 h)	7	14	18
Plunger pump	159, 331, 444, 476, 636, 671, 725, 849, 863, 1037, 1076, 1139, 1184, 1259, 1315, 1385, 1441, 1461, 1476, 1486	750 h (SPM 200 h)	7	13	-

overhaul cycle, given in Table 3.1. If we consider the overhaul as perfect, then the following point emerges:

If the overhaul renders the system to the AGAN state, then why is there a rise in the number of failures after each successive overhaul?

One of the most plausible explanations for the above conditions could be an "imperfect overhaul." During the overhaul, all efforts are made to rejuvenate the system, but all the components are not replaced, and some of them are subjected to repair activities as well. Moreover, the aging effect also hinders the system to get restored to the AGAN state. Overhauling could bring the system closer to the AGAN state (not exactly at the AGAN condition) depending on the resources available during the overhaul activity. In the literature, the available Kijima and Nasr *et al.* models consider CM and SPM as imperfect, respectively, without considering the overhaul activity separately.

The assumption of overhaul might lead to the underestimation of the parameters in real-life circumstances. Hence, in performance parameter estimation, it is very essential to treat the overhaul as imperfect to achieve more realistic solutions.

3.3 Imperfect Overhaul Virtual Age Model (IOVAM)

Scheduled preventive maintenance and overhaul activities differ from each other in several aspects as discussed earlier. Thus, quality indices for SPM and overhaul may be treated separately for a system undergoing MRO. A Kijima I-based IOVAM, which considers SPM, CM, and overhaul as imperfect, is shown in Eq. (3.1) (Sharma and Rai, 2020b). The concept of IOVAM is shown in Figure 3.1.

The model treats SPM and overhaul as a part of PM. Here, q_{CM} and q_{SPM} restore the damage that occurs at the time between the ith and $(i-1)^{th}$ intervention (i.e., CM or SPM), and q_O restores the damage that occurs at the time between the mth and $(m-1)^{th}$ overhaul.

$$V_{li} = (V_{l,i-1} + q_{CM}x_{li})^{\delta_{li}} \left[(V_{l,i-1} + q_{SPM}x_{li})^{1-h_{li}} (V_{l,m-1} + q_O X_{lm})^{h_i} \right]^{1-\delta_{li}} \tag{3.1}$$

where V_{li} is the virtual age after the ith intervention (CM or SPM or overhaul) of the lth system, x_{li} is the time between interventions (i.e., CM or

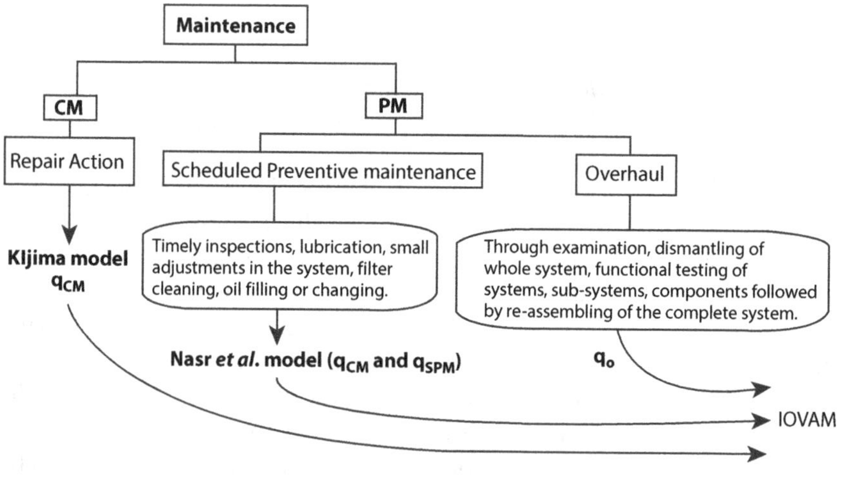

Figure 3.1 Imperfect CM, SPM, and overhaul in the IOVAM model.

SPM), X_{li} is the time between two overhauls, q_{CM} and q_{SPM} are the quality indices for CM and SPM, respectively, as described in Section 2.3.3.1, and

$$q_O = \begin{cases} 0\ ; & \text{for perfect Overhaul (as good as new condition)} \\ 0 < q_O < 1; & \text{for imperfect Overhaul (intermediate state)} \\ 1\ ; & \text{for minimal Overhaul (as bad as old condition)} \end{cases}$$

$$\delta_{li} = \begin{cases} 1 \text{ if CM is performed} \\ 0 \text{ if SPM is performed} \\ 0 \text{ if Overhaul is performed} \end{cases} \qquad h_{li} = \begin{cases} 1 \text{ for overhaul} \\ 0 \text{ otherwise} \end{cases}$$

The model follows the basic assumptions of the Kijima models as discussed in Chapter 1 (Section 2.3.3.1.). Equation (3.1) is explained with the help of a hypothetical example as given below.

Example: Let us consider that in a repairable system eight interventions ($n_l = 8$) occur in a particular time period. It can be seen from Table 3.2 that an imperfect CM is done at the first, second, third, fifth, sixth, and eighth interventions, and an imperfect SPM is accomplished at the fourth intervention. In addition, the seventh intervention is done to perform an imperfect overhaul. With the help of the IOVAM, the effect of q_{CM}, q_{SPM}, and q_O on system's virtual age can be appreciated.

Furthermore, the likelihood and log-likelihood functions are for the model as given in Eqs. (3.2) and (3.3).

$$L(x_1, x_2, \ldots\ldots x_n | \beta, \eta, q_{CM}, q_{PM}, q_O) =$$
$$\Pi_{l=1}^{K}\left[\Pi_{i=1}^{n_l}\left[\left[\left\{\frac{\beta}{\eta}\left(\frac{V_{l,i-1}+x_{li}}{\eta}\right)^{\beta-1}\right\}^{\delta_{li}} exp\left\{\left(\frac{V_{l,i-1}}{\eta}\right)^{\beta}-\left(\frac{V_{l,i-1}+x_{li}}{\eta}\right)^{\beta}\right\}\right]^{(1-h_{li})}\right.\right.$$
$$\left.\left.\left[\Pi_{m=1}^{U_l} exp\left\{\left(\frac{V_{l,m-1}}{\eta}\right)^{\beta}-\left(\frac{V_{l,m-1}+x_{lm}}{\eta}\right)^{\beta}\right\}\right]^{(h_{li})}\right]\right] \tag{3.2}$$

Table 3.2 Virtual age estimation using IOVAM with imperfect SPM, CM, and overhaul.

Intervention (i)	**Time of intervention (since new)**	**Time between interventions (TBI) (h)**	δ_{li}	h_{li}	V_{li}
1	t_{11} (CM)	$x_{11} = t_{11} - 0$	1	0	$V_1 = q_{CM}x_1$
2	t_{12} (CM)	$x_{12} = t_{12} - t_{11}$	1	0	$V_2 = V_1 + q_{CM}x_2$
3	t_{13} (CM)	$x_{13} = t_{13} - t_{12}$	1	0	$V_3 = V_2 + q_{CM}x_3$
4	t_{14} (SPM)	$x_{14} = t_{14} - t_{13}$	0	0	$V_4 = V_3 + qSPMx_4$
5	t_{15} (CM)	$x_{15} = t_{15} - t_{14}$	1	0	$V_5 = V_4 + q_{CM}x_5$
6	t_{16} (CM)	$x_{16} = t_{16} - t_{15}$	1	0	$V_6 = V_5 + q_{CM}x_6$
7	t_{17} (overhaul)	$X_m = t_{17} - t_{\text{at last overhaul}}$	0	1	$V_7 = V_{m-1} + q_O X_m$
8	t_{18} (CM)	$x_{18} = t_{18} - t_{17}$	1	0	$V_8 = V_7 + q_{CM}x_8$

Taking log on both sides

$$log(L) = log(\beta)\textstyle\sum_{l=1}^{K}[\sum_{i=1}^{n_l}(\delta_{li} - \delta_{li}h_{li})] + (\beta - 1)\sum_{l=1}^{K}[\sum_{i=1}^{n_l}(\delta_{li} - \delta_{li}h_{li})log(V_{l,i-1} + x_{li})] - \beta log(\eta)\sum_{l=1}^{K}[\sum_{i=1}^{n_l}(\delta_{li} - \delta_{li}h_{li})]$$
$$+\textstyle\sum_{l=1}^{K}\left[\sum_{i=1}^{n_l}(1-h_{li})\left[\left(\frac{V_{l,i-1}}{\eta}\right)^{\beta} - \left(\frac{(V_{l,i-1}+x_{li})}{\eta}\right)^{\beta}\right]\right]$$
$$+\textstyle\sum_{l=1}^{K}\left[\sum_{i=1}^{n_l}(h_{li})\left[\sum_{m=1}^{U_l}\left\{\left(\frac{V_{l,m-1}}{\eta}\right)^{\beta} - \left(\frac{V_{l,m-1}+x_{lm}}{\eta}\right)^{\beta}\right\}\right]\right] \quad (3.3)$$

In, Eqs. (3.2) and (3.3), V_{li} is given by Eq. (3.1). The performance parameters can be estimated by maximizing Eq. (3.3). The MATLAB code is appended in Appendix C.

In addition, the generalized virtual age model can also be suitably modified to some special cases as shown in Figures 3.2 and 3.3, to cover all cases of imperfect repair reliability analysis including those presented earlier in the literature.

The supposition of a perfect overhaul delivers a virtual notion that the operations can be met without incurring additional cost. However, an imperfect overhaul consideration may provide a lower estimate of availability figure, but it is much closer to the actual value. Hence, to meet the operational commitments, the additional cost incurred is the

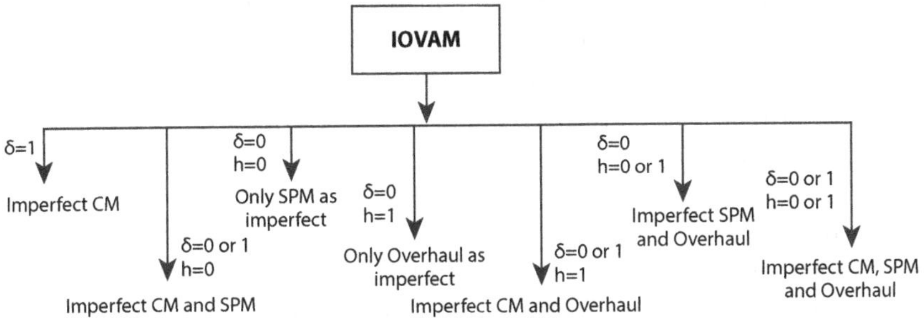

Figure 3.2 Modifications of the IOVAM.

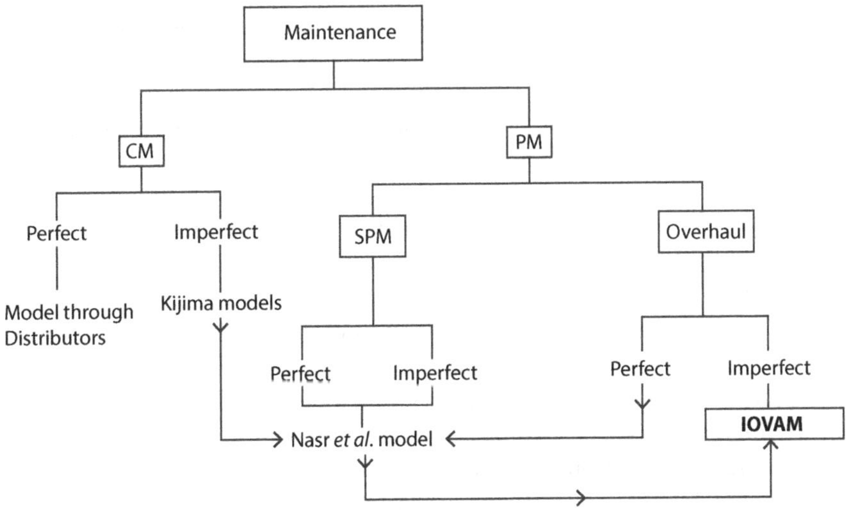

Figure 3.3 Incorporation of other models in the IOVAM.

cost that is actually required to be catered for, which actually cannot be visualized with perfect overhaul assumption. Thus, over a complete perspective, the actual cost does not increase when the overhaul is considered imperfect.

✓ Solved Example
As an example, let us consider the data sets of the turbo starter and plunger pump as described in Table 3.1. For the given data sets:

(f) Estimate the parameter considering the (i) IOVAM, (ii) Nasr *et al.* model, and (iii) NHPP model.
(g) Plot and compare the intensity and availability graphs obtained from the models mentioned in (a). Also, dictate the observations made from the analysis.
(h) Perform the sensitivity analysis considering the different values of q_o in IOVAM.

Solution:
(a) Performance parameters can be estimated with the help of:

i. IOVAM by maximizing Eq. (3.3) in MATLAB.
ii. NHPP model (using the MLEs as given in Appendix A (A1)) (Crow, 1975) and,
iii. Nasr *et al.* model (Nasr, Gasmi, and Sayadi, 2013) which considers only CM and SPM as imperfect (by maximizing Eq. (2.38) in MA TLAB considering the Nasr *et al.* KI model).

The results obtained for each data set are shown in Tables 3.3 and 3.4, respectively.

(b) A comparative analysis of the intensity function and the availability of both components assuming overhaul as perfect, imperfect, and minimal is shown in Figures 3.4–3.7.

The observations from the intensity function graphs of both components are as follows: (1) the intensity function becomes zero at the respective overhaul times when overhaul is assumed as perfect (Nasr *et al.* model) which indicates that components are brought to the AGAN condition; (2) the intensity function increases continuously when overhaul is assumed as minimal (NHPP model) which represents that even after the overhaul, the components assume the same condition as they were before the overhaul (ABAO condition); and (3) the intensity function of both components

Table 3.3 Reliability parameters estimation using different models for turbo starter (three overhaul cycle data).

Model	Parameters		
Nasr *et al.*	First overhaul cycle	q_{CM}	0.827
		q_{SPM}	0.750
		β	1.019
		η	215.754
	Second overhaul cycle	q_{CM}	0.664
		q_{SPM}	0.856
		β	1.203
		η	134.047
	Third overhaul cycle	q_{CM}	0.98
		q_{SPM}	0.87
		β	1.50
		η	114.145
NHPP	–	β	1.40
		η	136.99
IOVAM	–	q_{CM}	0.984000
		q_{SPM}	0.897000
		q_O	0.306000
		β	1.607000
		η	185.167

neither reaches zero nor at the same condition, when overhaul is assumed as imperfect, whereas it lies between the AGAN and the ABAO condition and increases gradually.

Likewise, availability graphs are also plotted for both components taking the mean repair time equal to 48 and 32 h for the turbo starter and plunger pump, respectively. When overhaul is assumed as perfect and minimal, the availability of both components reaches to one and decreases continuously,

Table 3.4 Reliability parameters estimation using different models for plunger pump (three overhaul cycle data).

Model	Parameters		
Nasr *et al.*	First overhaul cycle	q_{CM}	0.764
		q_{SPM}	0.852
		β	1.138
		η	146.871
	Second overhaul cycle	q_{CM}	0.881
		q_{SPM}	0.915
		β	1.389
		η	101.595
NHPP	–	β	1.49
		η	125.216
IOVAM	–	q_{CM}	0.973
		q_{SPM}	0.882
		q_O	0.128
		β	1.416
		η	112.636

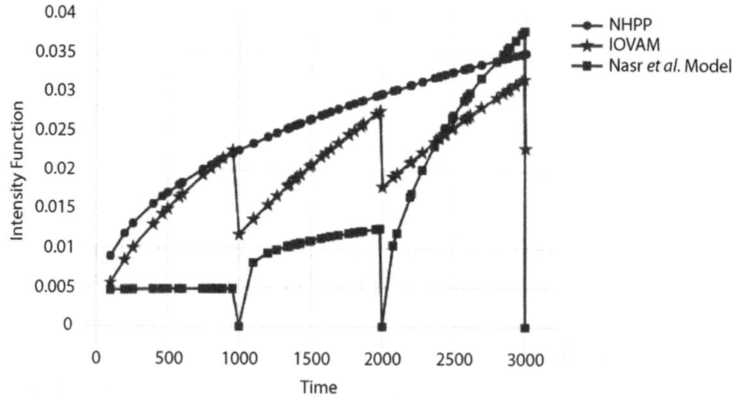

Figure 3.4 Intensity function graph of the turbo starter considering perfect (Nasr *et al.* model), imperfect (IOVAM), and minimal (NHPP) overhaul.

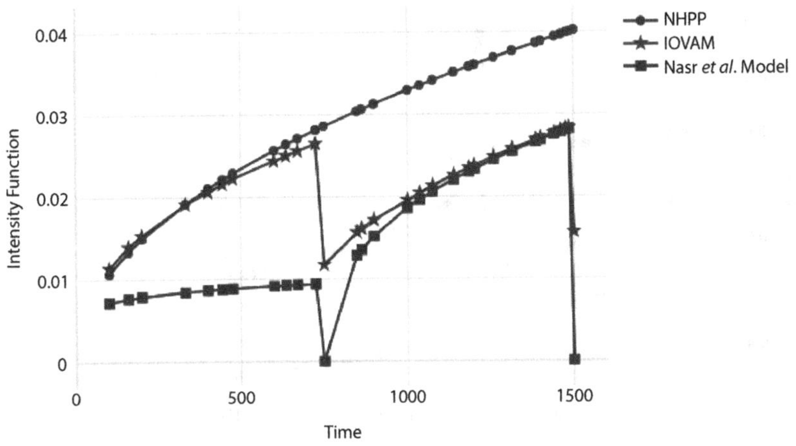

Figure 3.5 Intensity function graph of the plunger pump considering perfect (Nasr *et al.* model), imperfect (IOVAM), and minimal (NHPP) overhaul.

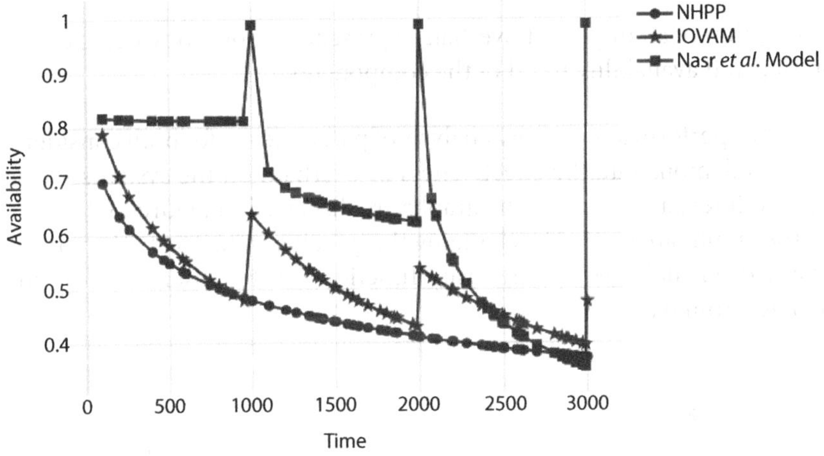

Figure 3.6 Availability graph of the turbo starter considering perfect (Nasr *et al.* model), imperfect (IOVAM), and minimal (NHPP) overhaul.

respectively. On the contrary, availability increases slightly and again decreases with time in case of an imperfect overhaul. Since the availability analysis of the component directly affects the cost analysis, it is very significant to examine the accurate trend of availability. The comparative analysis indicates that the theory of perfect overhaul overrates and minimal overhaul underrates the performance of the components, whereas the

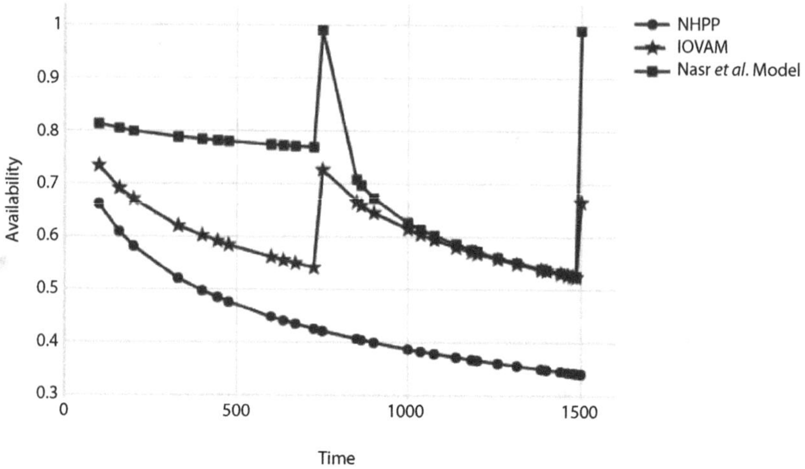

Figure 3.7 Availability graph of the plunger pump considering perfect (Nasr *et al.* model), imperfect (IOVAM), and minimal (NHPP) overhaul.

assumption of an imperfect overhaul represents a more convincing deteriorating and availability trend of the components.

(c) Let us perform a sensitivity analysis of parameter q_O for both considered cases to demonstrate the effectiveness of overhaul on the system's virtual age and intensity function. The graphs are displayed in Figures 3.8–3.11.

The graphs are plotted by changing the q_O value from 0 to 1. The renewal profile of virtual ages and intensity function can be observed as q_O that changes from 0 to 1.

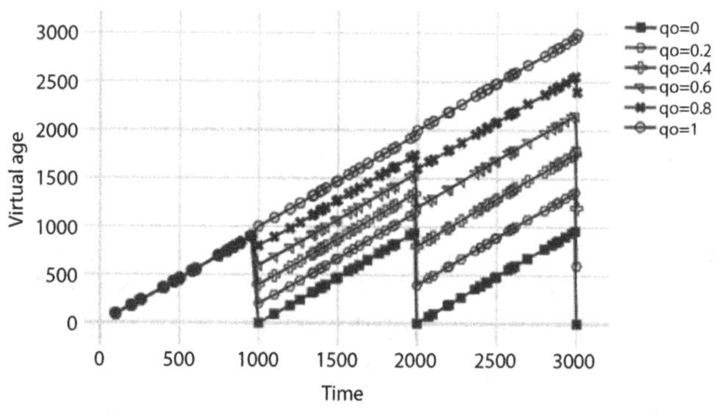

Figure 3.8 The turbo starter virtual age sensitivity graphs considering different values of q_O.

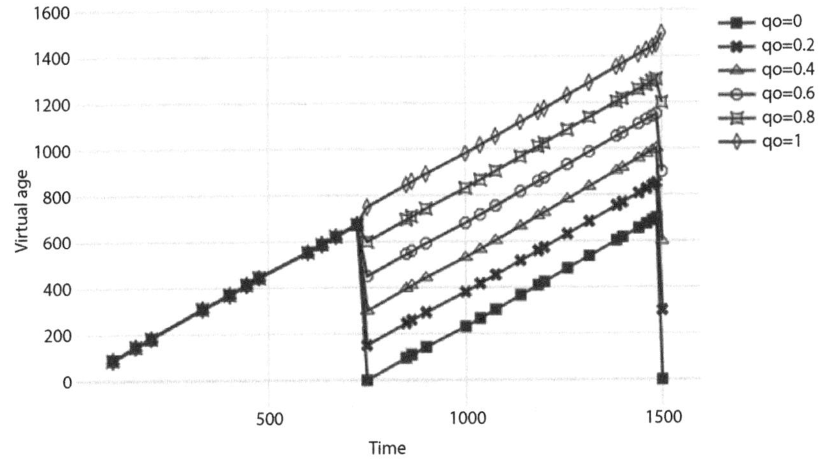

Figure 3.9 The plunger pump virtual age sensitivity graphs considering different values of q_O.

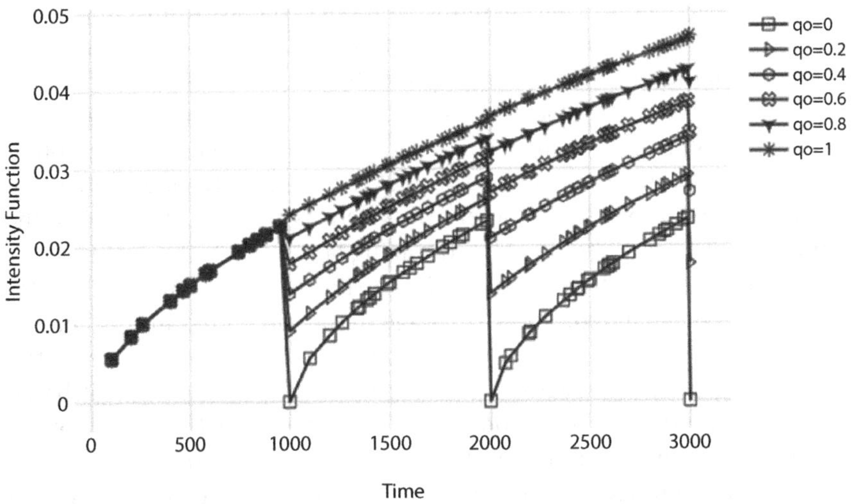

Figure 3.10 The turbo starter intensity function sensitivity graphs considering different values of q_O.

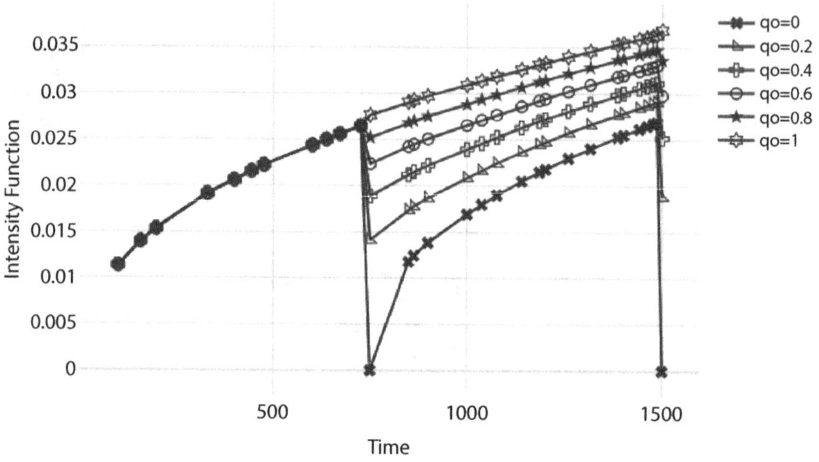

Figure 3.11 The plunger pump intensity function sensitivity graphs considering different values of q_O.

3.4 Chapter Summary

In this chapter, the concept of an imperfect overhaul is introduced which is mainly based on the concept of the KI model. The work aims to pay due consideration toward overhaul quality done on the system. The literature is more oriented toward optimizing overhaul intervals to minimize the total cost and normally considers SPM or CM as imperfect assuming that overhaul renders the system to the AGAN state. However, this assumption of a perfect overhaul may not be correct for complex repairable systems, since during the overhaul, the entire system is not replaced, rather the components undergo repair. Also, the aging factor affects the quality of the overhaul. Hence, with SPM and CM, the effectiveness of overhaul is also required to be analyzed. The virtual age model presented in this chapter works toward waiving the assumption of the AGAN state of a system after undergoing an overhaul activity. A new parameter q_O in the Kijima I-based virtual age model can be introduced, which represents the quality of the overhaul, and the likelihood and log-likelihood function can be derived to estimate the model parameters. The outcomes show that the theory of perfect and minimal overhaul overrates and underrates the performance of the components, respectively, whereas the assumption of an imperfect overhaul represents a more convincing deteriorating and availability trend of the components. The model and the obtained results are very useful for

Table 3.5 Aero engine failure times in hours.

Component name	Failure times (in hours)	Overhaul time
Aero engine (series alpha)	14.0667, 14.5667, 25.9167, 48.4500, 55.0333, 61.3000, 68.1833, 76.4333, 81.6333, 90.3000, 102.5667, 106.6167, 123.7000, 152.6000, 154.0667, 160.9000, 164.0500, 168.2167, 174.8500, 203.3833, 226.7667, 230.7833, 247.8667, 249.2500, 254.3500, 265.1000, 267.7333, 277.1167, 287.5167, 287.5500, 299.8667, 303.5167, 313.1667, 318.6167, 319.8655, 321.7000, 324.9333, 325.7500, 326.0500, 342.9500, 349.2000, 352.0667, 394.5833, 414.2000, 426.8500, 448.5042, 451.4167, 531.8667, 536.8667, 578.9167, 585.8833, 588.5000, 630.1333, 631.2333, 641.3833, 648.2333, 661.2667, 667.5500, 697.5167, 701.4667, 733.5167, 742.6667, 750.4833, 789.3667, 792.6667, 818.5500, 829.2333, 851.6667, 865.0667, 899.3167, 964.8667, 966.4667, 975.9667, 1041.0833, 1067.1833, 1268.2167, 1291.3500	550 h

MRO facilities while analyzing the repairable system effectiveness for reliability improvement.

❖ Exercise Problem

Aero engines used in military aircraft are empowered with high thrust to enable sudden climb and sustain high "G" loads during maneuvers. They are also designed to prevent surge and stall due to back pressure resulting from firing of rockets and missiles that create abundance of turbulence in front of the aero engine. These engines are subjected to very high aerodynamic and thermal stresses, hence are subjected to frequent failures.

The failure times in hours of such an aero engine for three overhaul cycles with time between overhauls of 550 h are given in Table 3.5. The SPM is done at 100 h.

(a) Estimate the performance parameters β, η, q_{CM}, q_{SPM}, and q_O for the given data set.
(b) Compare the results with the Kijima and NHPP models.
(c) Perform the sensitivity analysis considering the different values of q_O.
(d) Plot the intensity and availability graphs for each overhaul cycle and analyze the results.

References

Crow, L.H., Reliability analysis for complex, repairable systems, ARMY MATERIEL SYSTEMS ANALYSIS ACTIVITYABERDEEN PROVING GROUND MD, *Reliability and Biometry*, 13, 6, 379–410, 1975.

Nasr, A., Gasmi, S., Sayadi, M., Estimation of the parameters for a complex repairable system with preventive and corrective maintenance, in: *International Conference On Electrical Engineering and Software Applications (ICEESA)*, pp. 1–6, 2013.

Pascual, R. and Ortega, J.H., Optimal replacement and overhaul decisions with imperfect maintenance and warranty contracts. *Selected Papers Presented at QUALITA 2003*, vol. 91, pp. 241–48, 2006, https://doi.org/10.1016/j.ress.2005.01.018.

Seo, J.H. and Bai, D.S., An optimal maintenance policy for a system under periodic overhaul. *Math. Comput. Model.*, 39, 4–5, 373–80, 2004.

Sharma, G. and Rai, R.N., Reliability parameter estimation of repairable systems with imperfect maintenance, repair and overhaul. *Int. J. Qual. Reliab. Manag.*, 38, 4, 892–907, 2020b.

4

Techniques for Modeling and Analysis of Censored Data Considering the FM Approach

4.1 Introduction

Reliability analysis of repairable systems has always been a challenging task for industries, especially when the collected data are censored. In real-world situations, MRO facilities in the industries generally preserve all the system's related information such as the number of failures, time between failures, and their respective FMs for future reliability analysis. The problem arises when FM-wise analysis is required to be done for repairable systems along with other censoring criteria such as test termination and removing the operating system from the stud. The literature provides various models to deal with different types of censored data but lacks in providing FM-based censored data analysis techniques for repairable systems when both SPM and CM are treated as imperfect. Thus, a technique and virtual age models for repairable systems are required that can simultaneously address FM-wise analysis with right and multiply censoring data considering both the SPM and CM as imperfect.

The generalized renewal process (GRP) can be used to model all possible after-repair conditions whether it is AGAN, or in minimal or intermediate condition, and hence can provide more reliable results for repairable systems. The GRP-based general likelihood equation (Vassiliou, 2000) considering censored data is given by:

$$L = \prod_{i \in Failure} f(r_i|\beta,\eta,q) \prod_{i \in Censor} [1 - F(r_i^{+}|\beta,\eta,q)] \tag{4.1}$$

where r_i^{+} is the censored time.

Garima Sharma and Rajiv Nandan Rai. Advanced Techniques for Maintenance Modeling and Reliability Analysis of Repairable Systems, (81–110) © 2023 Scrivener Publishing LLC

It can be noted that the main difference in forming a methodology to deal with censored data analysis in case of repairable systems as compared with the nonrepairable systems lies in accommodating virtual age, which incorporates the effect of repair actions in it. Thus, real time is converted into a virtual scale depending on the quality of the repair. Hence, estimation of virtual age becomes imperative and challenging while carrying out reliability analysis of repairable systems, when data contain censored times.

The MLE-based parametric method using Eq. (4.1) is already being used to deal with censored data of repairable systems in case where all failures are considered irrespective of the concerned FMs (Vassiliou, 2000; Meyer and Krueger, 2001). In such analysis, all the data entries are considered as failure times until test termination or system removal time occurs as explained in Figure 4.1. Such an approach is referred to as the "black box (BB) approach." In other words, the available Kijima (GRP)-based maximum likelihood estimation method for censored data uses all failure times irrespective of their FMs.

However, a more precise failure analysis can be performed by identifying and segregating various FMs from the failure data (Sharma and Rai, 2018a; Sharma and Rai, 2021e). This is termed the "FM" approach. *The failure mode approach-based censored data analysis considers failure times of a particular FM (to be analyzed), keeping the failure times of all other FMs as censored* (Vassiliou, 2000). Presently, the parametric methods to deal with

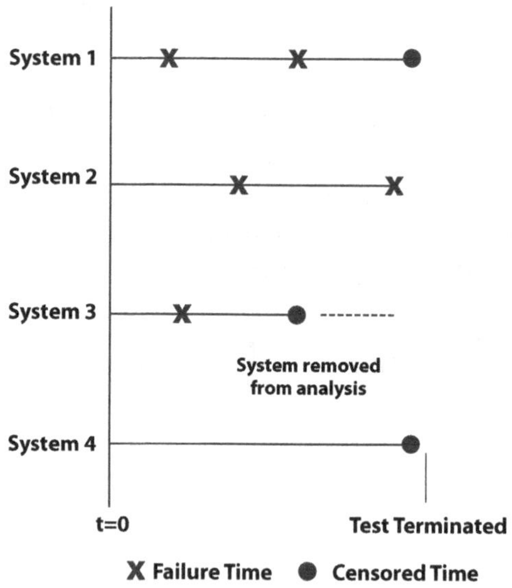

Figure 4.1 The black box approach for censored data.

censored data using the FM approach are available only for nonrepairable systems, which are generally categorized under multiply censoring, but the problem of censored data analysis for repairable systems considering different FMs has not yet been addressed in the literature. In addition, the analysis becomes even more complicated if SPM done on the repairable systems is also considered imperfect along with an imperfect CM.

To overcome the abovementioned challenges and limitations, a technique using virtual age models, which can simultaneously address FM-wise analysis with right and multiply censoring data, considering both the SPM and CM as imperfect, is required for repairable systems. In this chapter, we discuss the technique and the models to deal with it.

The chapter also provides an industrial perspective of the technique and the models by highlighting their applicability in conducting reliability analysis of complex and critical repairable systems. Conversion of the models to other GRP-based existing models is also presented.

4.2 Problem Background

Repairable systems could fail more than once due to the same or different FMs and are put back into operation after they are subjected to an imperfect repair. Hence, on considering the FM approach of data censoring, a system may get censored multiple times as shown in Figure 4.2(a) (considering FM1 as interested FM). The analysis acquires more complicacy when the FM approach is required to be analyzed in conjunction with the imperfect PM, as shown in Figure 4.2(b). For such a situation, a problem statement would be:

Problem Statement: A repairable system can fail due to various FMs. The performance parameters are required to be estimated for a particular FM (say FM1) considering both SPM and CM as imperfect and keeping other uninterested FMs as censored for the cases wherein:

Case I: The test is terminated at a fixed time, i.e., in case of type I censoring.

Case II: The test is terminated after a fixed number of failures, i.e., in case of type II censoring.

Case III: The systems are running but censoring occurs at many different times, i.e., in case of multiply censoring.

The removal of systems due to some environmental conditions also occurs during the observation period.

Considering the abovementioned problem statement, the basic terminologies and methodology are described in the subsequent sections.

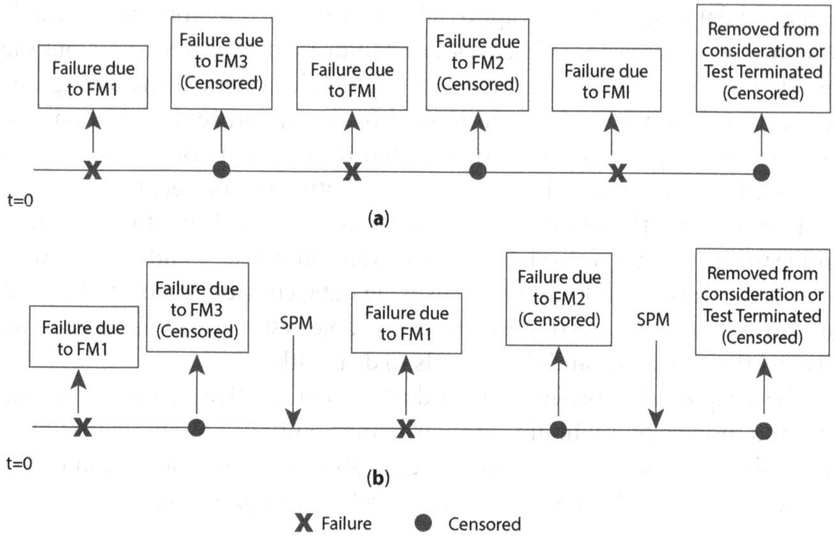

Figure 4.2 (a) The failure mode approach for repairable systems; (b) the failure mode approach for repairable systems with an imperfect SPM.

4.3 Basic Terminologies

Intervention: Any activity, which sojourns the system to perform its intended operation, is called intervention, and the time, when the intervention happens, is termed the intervention time.

Now, suppose there are K systems with $l = (1,2,3, \ldots. K)$, starting at $t = 0$. Let $0 \leq T_{l1} < T_{l2} < \cdots T_{Kn}$ symbolize the arrival time, where T_{li} is the arrival time of ith intervention of the lth system, and n is the number of interventions. Similarly, let X_{li} be the interarrival time between the ith and $(i - 1)^{th}$ intervention of the lth system. Furthermore, let t_i be the observed time of the ith intervention and x_i be the observed interarrival time between the ith and $(i - 1)^{th}$ intervention $(x_i = t_i - t_{i-1})$. Hence, t_{li} and x_{li} will be the observed time of the ith intervention and the observed interarrival time between the ith and $(i - 1)^{th}$ intervention of the lth system, respectively.

Primarily, two types of interventions can occur:

1. *Uncensored Intervention:* An intervention is "uncensored" when:
   It happened due to an interested FM.
   It is done to perform SPM

Thus, the corresponding time of the uncensored intervention will be "observed uncensored intervention time."

2. *Censored Intervention:* An intervention is "censored" when:
   It is done to remove the system from the analysis.
   The test was terminated.
   It happened due to uninterested FMs.
   Thus, the corresponding time of the censored intervention will be the "observed censored intervention time." Let us name the "observed censored intervention time" as "C_{TIME}" and denote it as t^{+}_{li}. Now, if the ith intervention is a censored intervention, then the observed interarrival time between the ith and $(i-1)^{th}$ intervention (which could be any uncensored or censored intervention) will be denoted as x^{+}_{li}. The concept of C_{TIME} assumes that the system has not failed at that point in time.

For a better comprehension of terminologies, suppose a system is operational from $t = 0$ h and undergoes three CMs at t_{11}, t_{12}, and t_{14} due to interested FM along with an SPM at t_{13} and the intervention time gets censored at t_{15} because of test termination as shown in Figure 4.3.

In this case, the terminologies will be as follows:

$$\Rightarrow x_{11} = t_{11} - 0$$

$$\Rightarrow x_{12} = t_{12} - t_{11}$$

$$\Rightarrow x_{13} = t_{13}(SPM) - t_{12}$$

$$\Rightarrow x_{14} = t_{14} - t_{13}$$

$$\Rightarrow x_{15}^{+} = t_{15}^{+} - t_{14} = C_{TIME} - t_{14}$$

In Figure 4.3, the circle with a cross represents t_{li} due to interested FM (uncensored interventions), the empty circle represents t_{li} due to SPM

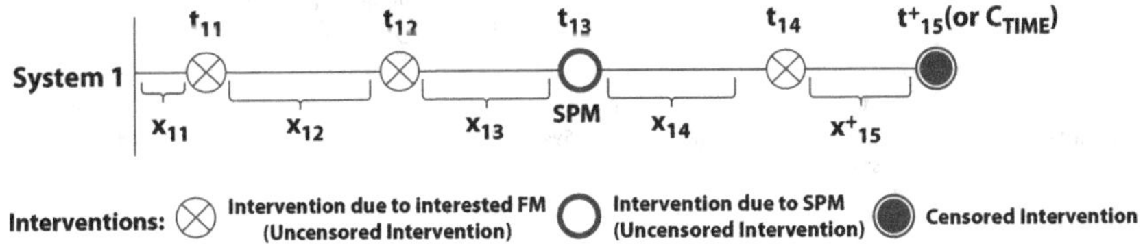

Figure 4.3 Notations for various intervention times.

(uncensored interventions), and the solid circles represent C_{TIMEs}. From now onward, the same notations and symbols of different interventions are used throughout the chapter.

4.4 Representation and Analysis of Cases

Using the terminologies as discussed in the previous section, let us now describe the cases stated in Section 4.2 and compare them with the BB approach for a better understanding.

Let us assume four identical repairable systems, i.e., 1, 2, 3, and 4. For each repairable system, an instant imperfect repair is done after failure besides the imperfect SPMs at scheduled times. Each case is explained with the help of figures and tables explained subsequently. To explain the FM approach, three FMs (FM1, FM2, and FM3) are considered with FM1 as an interested FM for analysis.

4.4.1 Case I

The systems are observed from $t = 0$ h till time say $t = T$ hours. Suppose that during the test, systems 1 and 2 fail twice with one SPM intervention in the duration of 0 to T hours. System 3 fails once and is then removed from the test. System 4 does not fail even once till time T, and only one SPM is performed on the system.

The representation of this case is shown in Figure 4.4(a) and (b) for the BB and FM approaches, respectively. Also, the C_{TIMEs} and x^{+}_{li} are shown in Table 4.1 for the BB and FM approaches.

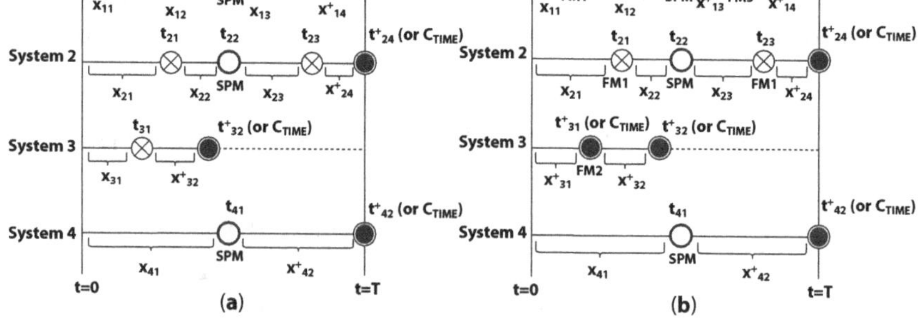

Figure 4.4 Representation of case I: (a) black box approach and (b) failure mode approach.

Table 4.1 Representation of censored data for case I.

Approach	Black box approach		Failure mode approach considering FM1	
System no.	t_{li}	x_{li}	t_{li}	x_{li}
System 1	t_{11}	x_{11}	t_{11} $(FM1)$	x_{11}
	t_{12} (SPM)	x_{12}	t_{12} (SPM)	x_{12}
	t_{13}	x_{13}	$t_{13}^{+}(FM3)$	x_{13}^{+}
	$t_{14}^{+}=T$	x_{14}^{+}	$t_{14}^{+}=T$	x_{14}^{+}
System 2	t_{21}	x_{21}	t_{21} $(FM1)$	x_{21}
	t_{22} (SPM)	x_{22}	t_{22} (SPM)	x_{22}
	t_{23}	x_{23}	t_{23} $(FM1)$	x_{23}
	$t_{24}^{+}=T$	x_{24}^{+}	$t_{24}^{+}=T$	x_{24}^{+}
System 3	t_{31}	x_{31}	$t_{31}^{+}(FM2)$	x_{31}^{+}
	t_{32}^{+}	x_{32}^{+}	t_{32}^{+}	x_{32}^{+}
System 4	t_{41} (SPM)	x_{41}	t_{41} (SPM)	x_{41}
	$t_{42}^{+}=T$	x_{42}^{+}	$t_{42}^{+}=T$	x_{42}^{+}

It can be observed from Figure 4.4 and Table 4.1 that there are four C_{TIMEs} (in hours), i.e., $t^{+}_{14}, t^{+}_{24}, t^{+}_{32}$, and t^{+}_{42}, when the BB approach is considered. But there are two additional C_{TIMEs}, i.e., t^{+}_{13} and t^{+}_{31}, when the FM approach is considered to analyze the system with respect to FM1, since intervention time because of other FMs (i.e., FM2 and FM3) are treated as censored.

4.4.2 Case II

The systems are observed from $t = 0$ h and the test is terminated say after the third failure. Now, suppose during the test, the first failure is observed in system 1, the second in system 2, and the last failure is observed again in system 1, and the test is terminated. One SPM intervention is performed in system 1 and system 2. System 3 is removed from the test without failing. System 4 does not fail even once, and only one SPM is performed on the system.

The modeling of the situation for the BB approach is presented in Figure 4.5, and C_{TIME} and x^{+}_{li} are shown in Table 4.2. It can be observed from Figure 4.5 and Table 4.2 that there are three C_{TIMEs}, i.e., t^{+}_{23}, t^{+}_{31}, and t^{+}_{42} when the BB approach is considered.

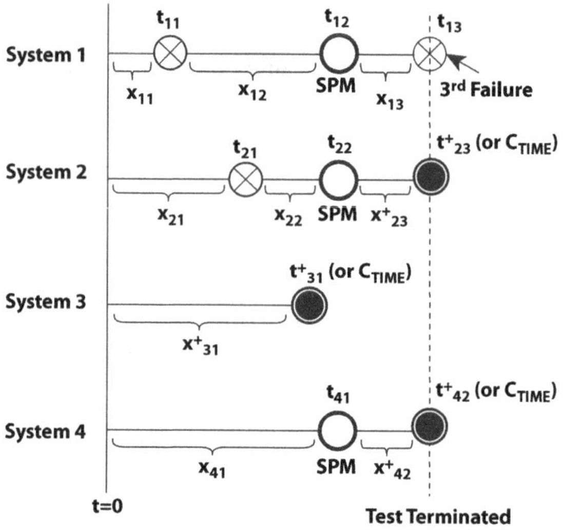

Figure 4.5 Representation of case II: the black box approach.

Table 4.2 Representation of censored data for case II: the black box approach.

Approach	**Black box approach**	
System no.	t_{li}	x_{li}
System 1	t_{11}	x_{11}
	t_{12} (*SPM*)	x_{12}
	t_{13}	x_{13}
System 2	t_{21}	x_{21}
	t_{22} (*SPM*)	x_{22}
	t_{23}^{+}	x_{23}^{+}
System 3	t_{31}^{+}	x_{31}^{+}
System 4	t_{41} (*SPM*)	x_{41}
	t_{42}^{+}	x_{42}^{+}

However, for the FM approach as shown in Figure 4.6, there could be two situations: (1) consider any three failures of any FM (Figure 4.6(a)) and (2) consider three failures of interested FM only (Figure 4.6(b)). In the first situation, there are three failures and three censored times (the same as the BB approach), but in the second situation, there are four failures in

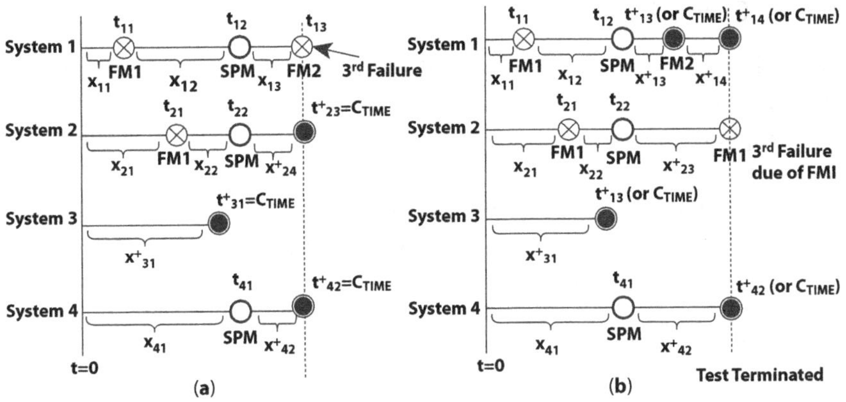

Figure 4.6 Representation of case II: the failure mode approach (a) considering any three failures of any FM and (b) considering three failures of interested FM only.

total: three because of FM1 (interested FM) and one because of another FM, i.e., FM2. For the second situation, after the occurrence of the third failure because of FM1 in system 2, the test is terminated. Table 4.3 shows the C_{TIME} and x^{+}_{li} for both conditions.

Table 4.3 Representation of censored data for case II: the failure mode approach.

Approach	Failure mode approach considering FM1			
	Situation (a)		Situation (b)	
System no.	t_{li}	x_{li}	t_{li}	x_{li}
System 1	t_{11} (FM1)	x_{11}	t_{11} (FM1)	x_{11}
	t_{12} (SPM)	x_{12}	t_{12} (SPM)	x_{12}
	t_{13} (FM2)	x_{13}	t^{+}_{13}(FM2)	x^{+}_{13}
	–	–	t^{+}_{14}	x^{+}_{14}
System 2	t_{21} (FM1)	x_{21}	t_{21} (FM1)	x_{21}
	t_{22} (SPM)	x_{22}	t_{22} (SPM)	x_{22}
	t^{+}_{23}	x^{+}_{23}	t_{23} (FM1)	x_{23}
System 3	t^{+}_{31}	x^{+}_{31}	t^{+}_{31}	x^{+}_{31}
System 4	t_{41} (SPM)	x_{41}	t_{41} (SPM)	x_{41}
	t^{+}_{42}	x^{+}_{42}	t^{+}_{42}	x^{+}_{42}

4.4.3 Case III

Case III occurs when systems are still running and analysis is required to be done considering a particular failure mode.

To explain this case, let us consider four systems as explained earlier. At the time of observation, systems 1 and 2 failed twice with one SPM intervention, system 3 failed once and is removed from consideration, and system 4 did not fail even once with only one SPM performed on the system. Considering the BB approach, only system 3 will be treated as censored at $C_{TIME} = t^{+}_{32}$. However, while considering the FM approach w.r.t. FM1, C_{TIMEs} will be t^{+}_{13}, t^{+}_{21}, and t^{+}_{32}. Case III is represented in Figure 4.7 considering both the BB and FM approaches and is further explained with the help of Table 4.4.

4.5 Models for FM-Wise Censored Data Analysis for Repairable Systems

Having introduced the concept of the FM approach-based censoring of repairable systems in sections 4.3 and 4.4, this section presents two virtual age models, namely, model I and model II (Sharma and Rai, 2021b), to model the FM approach-based censored data considering both SPM and CM as imperfect. The models are based on the concept of the KI and KII models (Kijima, 1989) and follow the basic assumptions of the Kijima models as discussed in Section 2.3.3.1.

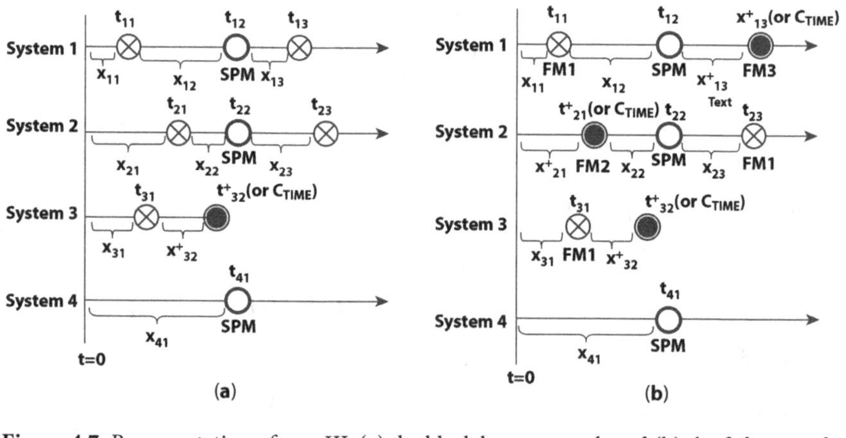

Figure 4.7 Representation of case III: (a) the black box approach and (b) the failure mode approach.

Table 4.4 Representation of censored data for case III.

Approach	**Black box approach**		**Failure mode approach considering FM1**	
System no.	t_{li}	x_{li}	t_{li}	x_{li}
System 1	t_{11}	x_{11}	t_{11} *(FM1)*	x_{11}
	t_{12} *(SPM)*	x_{12}	t_{12} *(SPM)*	x_{12}
	t_{13}	x_{13}	$t_{13}^{+}(FM3)$	x_{13}^{+}
System 2	t_{21}	x_{21}	$t_{21}^{+}(FM2)$	x_{21}^{+}
	t_{22} *(SPM)*	x_{22}	t_{22} *(SPM)*	x_{22}
	t_{23}	x_{23}	t_{23} *(FM1)*	x_{23}
System 3	t_{31}	x_{31}	t_{31} *(FM1)*	x_{31}
	t_{32}^{+}	x_{32}^{+}	t_{32}^{+}	x_{32}^{+}
System 4	t_{41} *(SPM)*	x_{41}	t_{41} *(SPM)*	x_{41}

4.5.1 Model I

The ith intervention would only compensate for the damage incurred after the $(i-1)^{th}$ intervention depending on its quality index, only when it is an uncensored intervention. Mathematically, the expression for virtual age considering the stated concept can be defined as:

$$\begin{aligned} V_{li} = V_{l,i-1} & \\ & + q_{CM}x_{li;}\ \textit{for intervention due to interested FM}, Vl, i-1 \\ & + qSPMx_{li};\ \textit{for intervention due to interested SPM}, V_{l,i-1} \\ & + x^{+}{}_{li};\ \textit{for censored intervention.} \end{aligned} \tag{4.2}$$

where V_{li} is the virtual age given to the lth systems after the ith intervention with $V_0 = 0$; q_{CM} and q_{SPM} are the quality index for CM and SPM, respectively, and confined within [0,1].

With this notion, "model I" to estimate the virtual age at the ith intervention can be expressed as:

$$V_{li} = \sum_{j=1}^{i} \left(q_{CM}^{\delta_{lj}} q_{SPM}^{1-\delta_{lj}} \right)^{1-\phi_{lj}} x_{lj} \tag{4.3}$$

where

$$\varphi_{li} = \begin{pmatrix} 1 \; \textit{forcensored intervention time (i.e. } C_{\mathrm{TIME}}), \\ 0 \; \textit{for uncensored intervention time} \end{pmatrix} \tag{4.4}$$

and,

$$\delta_{li} = \begin{pmatrix} 1 \; \textit{when intervention is uncensored and} \\ \textit{due to interested FM(CM),} \\ 0 \; \textit{when intervention is uncensored} \\ \textit{and due to SPM,} \\ 0 \; \textit{when intervention is censored} \end{pmatrix} \tag{4.5}$$

The expanded form of model I will be:

$$\Rightarrow V_{l1} = \left(q_{CM}^{\delta_{l1}} q_{SPM}^{1-\delta_{l1}} \right)^{1-\varphi_{l1}} x_{l1}$$

$$\Rightarrow V_{l2} = \left(q_{CM}^{\delta_{l1}} q_{SPM}^{1-\delta_{l1}} \right)^{1-\varphi_{l1}} x_{l1} + \left(q_{CM}^{\delta_{l2}} q_{SPM}^{1-\delta_{l2}} \right)^{1-\varphi_{l2}} x_{l2}$$

$$\Rightarrow V_{l3} = \left(q_{CM}^{\delta_{l1}} q_{SPM}^{1-\delta_{l1}} \right)^{1-\varphi_{l1}} x_{l1} + \left(q_{CM}^{\delta_{l2}} q_{SPM}^{1-\delta_{l2}} \right)^{1-\varphi_{l2}} x_{l2} + \left(q_{CM}^{\delta_{l3}} q_{SPM}^{1-\delta_{l3}} \right)^{1-\varphi_{l3}} x_{l3}$$

$$\cdot\cdot$$

$$\Rightarrow V_{ln} = \left(q_{CM}^{\delta_{l1}} q_{SPM}^{1-\delta_{l1}} \right)^{1-\varphi_{l1}} x_{l1} + \left(q_{CM}^{\delta_{l2}} q_{SPM}^{1-\delta_{l2}} \right)^{1-\varphi_{l2}} x_{l2} + \cdots + \left(q_{CM}^{\delta_{ln}} q_{SPM}^{1-\delta_{ln}} \right)^{1-\varphi_{ln}} x_{ln}$$

4.5.2 Model II

The ith intervention could collectively remove the damages accumulated up to the present time depending on its quality index, only when it is an uncensored intervention. Mathematically, the expression of virtual age considering the stated concept can be defined as:

$$V_{li} = q_{CM}(V_{l,i-1} + x_{li});$$

$$\textit{for intervention due to interested FMq}\text{SPM } (V_{l,i-1} + x_{li});$$

$$\textit{for intervention due to SPM}(V_{l,i-1} + x^{+}_{li});$$

$$\textit{or censored intervention.} \tag{4.6}$$

With this notion, "model II" to estimate the virtual age at the ith intervention can be defined as:

$$V_{li} = \sum_{j=1}^{i} \left(q_{CM}^{(i-j+1-\zeta_{li})\delta_{lj}} q_{SPM}^{(i-j+1-\zeta_{li})(1-\delta_{lj})} \right)^{(1-\phi_{lj})} x_{lj} \tag{4.7}$$

where,

$$\zeta_{li} = \begin{cases} 0 \textit{ when intervention is uncensored,} \\ 1,2,3,..\textit{for consecutive censored} \\ \textit{intervention till next uncensored} \\ \textit{intervention.} \end{cases} \tag{4.8}$$

The concept of ζ_{li} can be explained as when there is uncensored intervention time, ζ_{li} value is 0. As the C_{TIME} appears, ζ_{li} is assigned the value of 1. In case of consecutive C_{TIMEs}, ζ_{li} assumes the value serially as 2, 3, 4, and so on. The value of ζ_{li} is again assigned 0 on the appearance of an uncensored intervention time and this cycle continues.

The expanded form of model II will be:

$$\Rightarrow V_{l1} = \left(q_{CM}^{(1-1+1-\zeta_{l1})\delta_{l1}} q_{SPM}^{(1-1+1-\zeta_{l1})(1-\delta_{l1})}\right)^{1-\varphi_{l1}} x_{l1}$$

$$\Rightarrow V_{l2} = \left(q_{CM}^{(2-1+1-\zeta_{l2})\delta_{l1}} q_{SPM}^{(2-1+1-\zeta_{l2})(1-\delta_{l1})}\right)^{1-\varphi_{l1}} x_{l1} + \left(q_{CM}^{(2-2+1-\zeta_{l2})\delta_{l2}} q_{SPM}^{(2-2+1-\zeta_{l2})(1-\delta_{l2})}\right)^{1-\varphi_{l2}} x_{l2}$$

$$\Rightarrow V_{l3} = \left(q_{CM}^{(3-1+1-\zeta_{l3})\delta_{l1}} q_{SPM}^{(3-1+1-\zeta_{l3})(1-\delta_{l1})}\right)^{1-\varphi_{l1}} x_{l1} + \left(q_{CM}^{(3-2+1-\zeta_{l3})\delta_{l2}} q_{SPM}^{(3-2+1-\zeta_{l3})(1-\delta_{l2})}\right)^{1-\varphi_{l2}} x_{l2} + \left(q_{CM}^{(3-3+1-\zeta_{l3})\delta_{l3}} q_{SPM}^{(3-3+1-\zeta_{l3})(1-\delta_{l3})}\right)^{1-\varphi_{l3}} x_{l3}$$

$$\vdots$$

$$\Rightarrow V_{ln} = \left(q_{CM}^{(n-1+1-\zeta_{ln})\delta_{l1}} q_{SPM}^{(n-1+1-\zeta_{ln})(1-\delta_{l1})}\right)^{1-\varphi_{l1}} x_{l1} + \left(q_{CM}^{(n-2+1-\zeta_{ln})\delta_{l2}} q_{SPM}^{(n-2+1-\zeta_{ln})(1-\delta_{l2})}\right)^{1-\varphi_{l2}} x_{l2} + \cdots + \left(q_{CM}^{(n-n+1-\zeta_{ln})\delta_{ln}} q_{SPM}^{(n-n+1-\zeta_{ln})(1-\delta_{ln})}\right)^{1-\varphi_{ln}} x_{ln}$$

To further illustrate the models, we consider a hypothetical case where a repairable system has failed seven times over a period because of three different FMs and is also subjected to a time-based imperfect SPM as shown in Figure 4.8. Now one wants to analyze the system due to failures inflicted by FM1. In that case, the first and fourth interventions are done due to FM1 and the sixth intervention is performed to carry out SPM. Thus, these

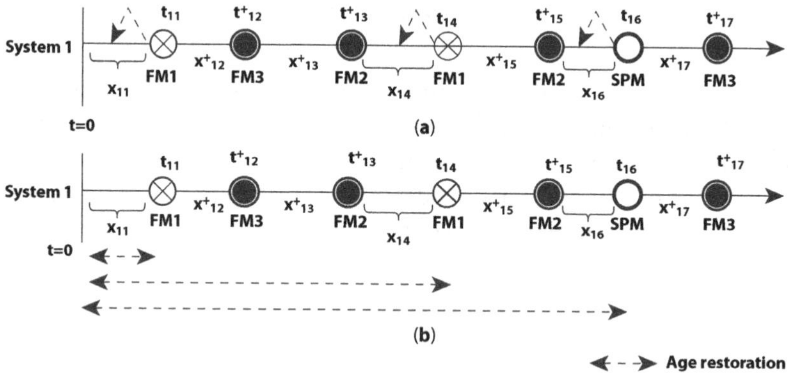

Figure 4.8 (a) Model I and (b) model II.

three intervention times are uncensored, whereas the second, third, fifth, and seventh intervention times are censored since the occurrence of failures is due to other FMs (i.e., either FM2 or FM3).

The next step is to assign the values of δ_{li}, φ_{li}, and ζ_{li} for various types of interventions. After assigning the values of δ_{li}, φ_{li}, and ζ_{li}, virtual ages are estimated using Eqs. (4.3) and (4.7) and further explained in Tables 4.5 and 4.6 for model I and model II, respectively.

It can be observed from Tables 4.5 and 4.6 that q_{CM} does not have any impact on C_{TIME} while estimating the virtual age. This is due to the assumption that at C_{TIME} the system has not failed or no maintenance action is performed. Thus, this intervention does not restore the life of the system at all. In contrast, at uncensored data points, the effect of qSPM and q_{CM} can be observed according to the maintenance actions performed at that point in time.

Furthermore, the likelihood and log-likelihood functions could be accordingly developed for parameter estimation (Ebeling, 2004). The likelihood and log-likelihood functions are given in Eqs. (4.9) and (4.10), respectively. MLEs can be estimated by taking derivatives of Eq. (4.10) with respect to β, η, q_{SPM}, and q_{CM} and equating them to 0. The equations for MLEs are provided in Appendix D.

However, the reliability parameters can also be estimated by maximizing the log-likelihood equation. Here, in Eqs. (4.9) and (4.10), V_{li} is given by Eqs. (4.3) and (4.7) for model I and model II, respectively, and n_l is the number of interventions in the lth system.

Table 4.5 Virtual age estimation for model I.

Intervention (i)	t_{li}	x_{li}	δ_{li}	φ_{li}	V_{li}
1	t_{11}	x_{11}	1	0	$q_{CM}x_{11}$
2	t_{12}^+	x_{12}^+	0	1	$q_{CM}x_{11}+x_{12}^+$
3	t_{13}^+	x_{13}^+	0	1	$q_{CM}x_{11}+x_{12}^++x_{13}^+$
4	t_{14}	x_{14}	1	0	$q_{CM}x_{11}+x_{12}^++x_{13}^++q_{CM}x_{14}$
5	t_{15}^+	x_{15}^+	0	1	$q_{CM}x_{11}+x_{12}^++x_{13}^++q_{CM}x_{14}+x_{15}^+$
6	t_{16}	x_{16}	0	0	$q_{CM}x_{11}+x_{12}^++x_{13}^++q_{CM}xl_{4}+x_{15}^++q_{SPM}x_{16}$
7	t_{17}^+	x_{17}^+	0	1	$q_{CM}x_{11}+x_{12}^++x_{13}^++q_{CM}x_{14}+x_{15}^+$ $+q_{SPM}x_{16}+x_{17}^+$

Table 4.6 Virtual age estimation for model II.

Intervention (i)	t_{li}	x_{li}	δ_{il}	φ_{li}	ζ_{li}	V_{li}
1	t_{11}	x_{11}	1	0	0	$q_{CM}x_{11}$
2	t_{12}^{+}	x_{12}^{+}	0	1	1	$q_{CM}x_{11}+x_{12}^{+}$
3	t_{13}^{+}	x_{13}^{+}	0	1	2	$q_{CM}x_{11}+x_{12}^{+}+x_{13}^{+}$
4	t_{14}	x_{14}	1	0	0	$q_{CM}{}^{4}x_{11}+x_{12}^{+}+x_{13}^{+}+q_{CM}x_{14}$
5	t_{15}^{+}	x_{15}^{+}	0	1	1	$q_{CM}{}^{4}x_{11}+x_{12}^{+}+x_{13}^{+}+q_{CM}x_{14}+x_{15}^{+}$
6	t_{16}	x_{16}	0	0	0	$q_{CM}{}^{6}x_{11}+x_{12}^{+}+x_{13}^{+}+q_{CM}{}^{3}x_{14}+x_{15}^{+}+q_{SPM}x_{16}$
7	t_{17}^{+}	x_{17}^{+}	0	1	1	$q_{CM}{}^{6}x_{11}+x_{12}^{+}+x_{13}^{+}+q_{CM}{}^{3}x_{14}+x_{15}^{+}+q_{SPM}x_{16}+x_{17}^{+}$

$$L(Data|\beta,\eta,q_{CM},q_{PM})=$$
$$\Pi_{l=1}^{K}\left[\Pi_{i=1}^{n_l}\left[\left[\left\{\frac{\beta}{\eta}\left(\frac{V_{l,i-1}+x_{li}}{\eta}\right)^{\beta-1}\right\}^{\delta_{li}} exp\left\{\left(\frac{V_{l,i-1}}{\eta}\right)^{\beta}\right.\right.\right.\right.$$
$$\left.\left.\left.\left.-\left(\frac{V_{l,i-1}+x_{li}}{\eta}\right)^{\beta}\right\}\right]^{(1-\varphi_{li})}\left[exp\left\{\left(\frac{V_{l,i-1}}{\eta}\right)^{\beta}-\left(\frac{V_{l,i-1}+x_{li}}{\eta}\right)^{\beta}\right\}\right]^{(\varphi_{li})}\right]\right] \tag{4.9}$$

Taking log on both sides

$$log(L)=log(\beta)\Sigma_{l=1}^{K}\left[\Sigma_{i=1}^{n_l}(\delta_{li}-\delta_{li}\varphi_{li})\right]+(\beta-1)\Sigma_{l=1}^{K}\left[\Sigma_{i=1}^{n_l}(\delta_{li}\right.$$
$$\left.-\delta_{li}\varphi_{li})log(V_{l,i-1}+x_{li})\right]-\beta log(\eta)\Sigma_{l=1}^{K}\left[\Sigma_{i=1}^{n_l}(\delta_{li}-\delta_{li}\varphi_{li})\right]$$
$$+\Sigma_{l=1}^{K}\left[\Sigma_{i=1}^{n_l}(1-\varphi_{l,i})\left[\left(\frac{V_{l,i-1}}{\eta}\right)^{\beta}-\left(\frac{(V_{l,i-1}+x_{li})}{\eta}\right)^{\beta}\right]\right]$$
$$+\Sigma_{l=1}^{K}\left[\Sigma_{i=1}^{n_l}(\varphi_{li})\left[\left(\frac{V_{l,i-1}}{\eta}\right)^{\beta}-\left(\frac{(V_{l,i-1}+x_{li})}{\eta}\right)^{\beta}\right]\right] \tag{4.10}$$

4.5.3 Comparison of FM-Wise Censoring Models with Kijima and Other Existing Models

The objective of FM-wise censoring models is to provide the solution to the problem which industries face while analyzing FM-wise censored data of repairable systems. The other models like NHPP (Crow, 1975), Kijima (Kijima, 1989), and Nasr *et al.* (Nasr, Gasmi, and Sayadi, 2013) can be effectively used for failure data analysis, but when the question arises on segregation of failure data w.r.t. their FM which involves censoring of failure data, the Kijima and other existing models lack in modeling and providing solutions in such cases. As discussed earlier, Kijima models can be modified to RP and NHPP depending on the value of q. Likewise, the Nasr *et al.* models can be modified to Kijima models. However, both models deal with failure data without considering the FM of the respective failure. The FM-wise censoring models are the superset of all models and can be converted into the NHPP, Kijima, and Nasr *et al.* models. The conversion of the models into existing ones is explained in Table 4.7.

For a better understanding of the readers, the applicability of FM-wise censored models to deal with different conditions such as imperfect CM, imperfect SPM, and censored data, as compared with the existing models in this domain is summarized in Table 4.8. The table reveals that basic KI and KII virtual age equations incorporate the effect of only imperfect CM. They do not consider the outcomes due to imperfect PM and data censoring. Nasr *et al.* (2013) effectively utilize the effect of imperfect SPM and CM in their virtual age equations but did not include the data censoring concept. Hence, the FM-wise censored models incorporate all the conditions in their virtual age equations by waiving the existing assumptions in the literature.

Table 4.7 Conversion of FM-wise censored models into existing ones.

Condition	Converted to
Put $\varphi_{li} = \zeta_{li} = 0$ and $\delta_{li} = 1$	Kijima models
Put $\varphi_{li} = \zeta_{li} = 0$ and $\delta_{li} = 1$ (for CM), $\delta_{li} = 0$ (for SPM)	Nasr *et al.* models

Table 4.8 Comparison matrix for existing and FM-wise censored models.

Virtual age model	Imperfect CM	Imperfect PM	Censoring concept
Existing Kijima I	A	NA	NA
Existing Kijima II	A	NA	NA
Existing Nasr *et al.*	A	A	NA
FM-wise censored models	A	A	A
A, available; NA, not available			

4.6 An Industrial Perspective of Technique

The industries dealing with large, complex, and critical repairable systems, for example, aerospace, chemical, or nuclear industries, generally preserve all the information such as the starting time of the system, the number of failures, time between failures, and their respective FMs for reliability analysis of the systems. But the industry still faces the practical challenges to analyze failure data due to different censoring criteria such as test termination, removing the operating system from the study during data collection, and at the same time, parameter estimation is required to be done considering various failure modes. This problem is mainly observed by the industries in right and multiply censoring because most of the time the data generated from reliability testing, warranty analysis, and field failures are generally right or multiply censored (Wu, 2012) in case of repairable systems.

The technique presented in this chapter will lead to a more precise estimation of reliability parameters that will help in a more accurate evaluation of scale and wear-out parameters along with correct estimation of probabilities of failure of various components and the equipment. This evaluation will further assist in conducting FMEA of critical repairable systems and will also help in an exact understanding of the remaining useful life of various components and systems. An effective FMEA and residual life estimation techniques will help industries in developing the most suitable maintenance policies that can sustain operations for a longer period of time with enhanced throughput. The applicability of the present methodology in conducting FMEA of repairable systems and estimation of residual life of various components is briefly described in the subsequent subparagraphs.

4.6.1 FMEA of Repairable Systems

During FMEA, industrial experts try to find out the root cause of failures and quantify the risk involved due to a particular FM using RPN. The RPN is given by:

$$RPN = O \times S \times D, \tag{4.11}$$

where O is the occurrence of failure due to a particular FM,
S is the severity involved due to FM, and
D is the level of detection to prevent the cause of that FM from occurring.

Here, the values of S and D are evaluated using a subjective scale of 1 to 10 with the help of field experts.

The literature provides various methods of RPN calculation based on different approaches (H.-C. Liu, Liu, and Liu, 2013) such as fuzzy sets (Hao Wang, Zhang, and Yang, 2019), Gaussian process (Bichon, McFarland, and Mahadevan, 2011), ELECTRE TRI (Certa *et al.*, 2017), general model (Kim and Zuo, 2018), weighted RPN (Xiao *et al.*, 2011), and logistic regression model (Bhattacharjee, Dey, and Mandal, 2020) to estimate RPN, but FMEA in case of repairable system usually gets affected because of mixed failure times of different failure modes. In addition, removing the uninterested FM failure times from analysis may affect the collected data as well as parameter estimation. In a such situation, estimating O becomes a very challenging task. The available methods could be very effective in estimating RPN, but lack in addressing the conceptual part of FM-wise censored data analysis in case of repairable systems. The methodology presented in this chapter mainly focuses on providing a parametric approach to estimate $F(t)$ for each FM separately which can be very helpful in assigning ranking for O to estimate FM as well as system-wise RPN for repairable systems. Having estimated O, one can estimate the RPN of repairable systems by adopting different approaches available in the literature. The subsequent subsection deliberates RPN estimation using the traditional method in which the occurrence (O) is assigned with the help of FM-wise censored models.

4.6.1.1 RPN Estimation Using FM-Wise Censored Models

To estimate O, the first thing is to find the probability of failure $F(.)$ of each FM. Since FMs are in series, the reliability of the complete system can be evaluated as explained below:

Suppose m FMs, FM_1, FM_2, FM_3, …FM_m have been worked out from the lifetime data of a particular component/system by an industry.

The model parameters of all the FMs followed by the probability of failure and reliability of individual FMs can then be estimated with the help of the methodology for FM-based data censoring technique for repairable systems.

Since the failure modes behave as a series model in reliability, the system reliability $R(.)$ is given by:

$$R\,(t)_{system} = R(t)_{FM1} \times R(t)_{FM2} \ldots \times R(t)_{FMm} \tag{4.12}$$

$$F\,(t)_{system} = 1 - [R(t)_{FM1} \times R(t)_{FM2} \ldots \times R(t)_{FMm}] \tag{4.13}$$

Once $F(.)$ is estimated, we can calculate O by developing a scale of 1–10 and assign these values to the concerned FMs depending on their probabilities of failures. The values of RPNs can be obtained for individual FMs and the complete system with the help of Eq. (4.11).

Hence, a complete investigation of all FMs and their parameter estimation would help in arriving at the most precise reliability figure of the complete system. It would be worthwhile bringing out that without the concept of FM-based data censoring for the repairable system as presented in this chapter, it would not have been possible to carry out a quantitative FM analysis for the repairable system.

Moreover, the models and methodology would be more helpful when the data sets are huge and contain too many FMs. The obtained parameters further can be used to improve the design of the system and develop maintenance policies accordingly.

4.6.2 Remaining Useful Life Estimation of Repairable Systems

Remaining useful life (RUL) estimation is the most common task in the research field for prognostics and health management. The issue of residual life estimation plays an important role in scheduling maintenance. It plays a key role in implementing condition-based maintenance decision-making. Moreover, by estimating the RUL of the target product, a frequently asked question in the industry about how long the target product can survive could be answered quantitatively.

The mean residual life (MRL) for repairable systems at any V_{li} shown in Eq. (4.14) is obtained by modifying the base MRL equation as given in:

$$MRL(V_{li}) = \frac{\int_{V_{li}}^{\infty} R(V)dV}{R(V_{li})}, \tag{4.14}$$

where V_{li} is given by Eqs. (4.3) and (4.7) for FM-wise censored model I and model II, respectively, and is considered a measure of RUL.

The values of the virtual age, obtained with the help of FM-based censored data analysis for repairable systems will provide a more accurate value of RUL of the repairable systems through which a more effective condition-based maintenance can be resorted to.

✓ Solved Example

To understand the presented technique and models in this chapter, let us consider a case from the aviation industry.

▪ Case and Data Description

The selected case is a turbojet engine with a twin-spool compressor mainly used in combat aircraft in military aviation. The aero engines are put back into operation after general repair. The aero engines undergo periodic overhauls; the original equipment manufacturer (OEM) specified the time between overhaul (TBO) being 550 operating hours. Within an overhaul cycle, the aero engines are subjected to OEM-specified periodic PMs at fixed intervals of 50 operating hours (i.e., at $SPM_1 = 50$, $SPM_2 = 100$, $SPM_3 = 150....SPM_{11} = 550$ h (overhaul)). The OEM-specified TTL of the aero engine is 1800 operating hours. The data set of three identical aero engines (multiple repairable systems ($K = 3$)) is provided. The system-wise failure times of all three systems are appended in Table 4.9.

Furthermore, nine FMs are observed and segregated from the failure data and are presented in Table 4.9 along with their failure times. It is observed that "compressor blades broken" (FM1), "oil leak" (FM2), and "overheat" (FM3) are the three most dominant FMs in this case. The industry wants to analyze the system deterioration due to the observed three most dominant FMs and estimate the model parameters for each FM one by one. Now, in view of the above data set:

(a) Identify the different types of interventions such as censored, uncensored, and SPM.
(b) Estimate the model parameters by maximizing Eq. (4.10) for both models considering each FM.
(c) Plot the intensity function graph with the help of estimated parameters and analyze the results.

Table 4.9 Data description of the three identical aero engines.

System-wise failure times (in operating hours)	
System 1	17, 39, 81, 119, 176, 197, 218, 237, 265, 277, 321, 332, 360, 384, 414, 423, 428, 443, 461, 474, 512, 521, 536, 545
System 2	29, 31, 68, 90, 128, 164, 261, 287, 325, 366, 395, 419, 444, 461, 490, 523, 532, 539, 545
System 3	22, 61, 88, 123, 171, 229, 278, 326, 344, 367, 382, 418, 438, 458, 476, 521, 544
Failure mode-wise segregated data sets (in operating hours)	
RPM fluctuation	90, 176, 229, 277, 532
Crack in diffuser casing	367, 444
Oil leak	39, 61, 81, 128, 261, 265, 278, 332, 366, 382, 419, 438, 461, 523
Overheat	17, 29, 123, 384, 418, 428, 458, 461, 476, 536, 539, 545
Compressor blades broken	22, 31, 88, 119, 218, 237, 287, 321, 325, 326, 344, 360, 395, 423, 443, 474, 490, 521, 544, 545
High oil consumption	68, 164, 414
Compressor bearing wear out	171, 521
Thrust variation	197
Hot gas leak	512

Solution:

(a) One overhaul cycle data is taken into consideration (test is terminated at 550 operating hours), and the data analysis takes the form of type I censoring with the FM approach (*case I*). The description of the data set now sorted for analysis is presented in Table 4.10 for the "compressor blades broken" failure mode. Similar data sets can be created for the other two dominant FMs. The data sets contain

Table 4.10 Data description of the three identical aero engines.

Data used for the analysis of compressor blades broken FM	
System 1 intervention times	17+, 39+, 50 (SPM), 81+, 100 (SPM), 119, 150 (SPM), 176+, 197+, 200 (SPM), 218, 237, 250 (SPM), 265+, 277+, 300 (SPM), 321, 332+, 350 (SPM), 360, 384+, 400 (SPM), 414+, 423, 428+, 443, 450 (SPM), 461+, 474, 500 (SPM), 512+, 521+, 536+, 545, 550+ (terminated)
System 2 intervention times	29+, 31, 50 (SPM), 68+, 90+, 100 (SPM), 128+, 150 (SPM), 164+, 200 (SPM), 250 (SPM), 261+, 287, 300 (SPM), 325, 350 (SPM), 366+, 395, 400 (SPM), 419+, 444+, 450 (SPM), 461+, 490, 500 (SPM), 523+, 532+, 539+, 545+, 550+ (terminated)
System 3 intervention times	22, 50 (SPM), 61+, 88, 100 (SPM), 123+, 150 (SPM), 171+, 200 (SPM), 229+, 250 (SPM), 278+, 300 (SPM), 326, 344, 350 (SPM), 367+, 382+, 400 (SPM), 418+, 438+, 450 (SPM), 458+, 476+, 500 (SPM), 521, 544, 550+ (terminated)

uncensored intervention times, i.e., failure times due to interested FM, SPM intervals of every 50 operating hours, and C_{TIMEs} due to uninterested FMs and test termination time. No system is removed from the analysis during the observation.

(b) Using the virtual age models, parameter estimation is carried out by maximizing the log-likelihood function as given in Eq. (4.10) (the code is appended in Appendix D). The obtained parameters are appended in Table 4.11.

The maximum log-likelihood value is obtained for model II for each dominant FM. This indicates that model II is more suitable for the considered data set selected for analysis than model I. If the interventions due to the same FM are considered, then the effect of maintenance action after each intervention is bound to affect the subsequent maintenance actions as prescribed by model II. In addition, the reason why the concept of model II is considered more appropriate for FM analysis is that the repair activities are similar after almost every intervention since a particular type of FM is considered at a time and similar repair

Table 4.11 Model parameters estimated from the models for the considered data set.

FMs	Model	Parameters	Values	LnL
FM1	Model I	q_{CM}	0.913	−107.58
		q_{SPM}	0.345	
		β	1.112	
		η	94.711	
	Model II	q_{CM}	0.401	−107.43
		q_{SPM}	0.531	
		β	1.043	
		η	83.949	
FM2	Model I	q_{CM}	0.438	−80.18
		q_{SPM}	0.399	
		β	1.036	
		η	118.53	
	Model II	q_{CM}	0.290	−80.14
		q_{SPM}	0.395	
		β	1.063	
		η	120.51	
FM3	Model I	q_{CM}	0.375	−70.71
		q_{SPM}	0.422	
		β	1.031	
		η	136.88	
	Model II	q_{CM}	0.214	−70.62
		q_{SPM}	0.669	
		β	1.011	
		η	133.52	

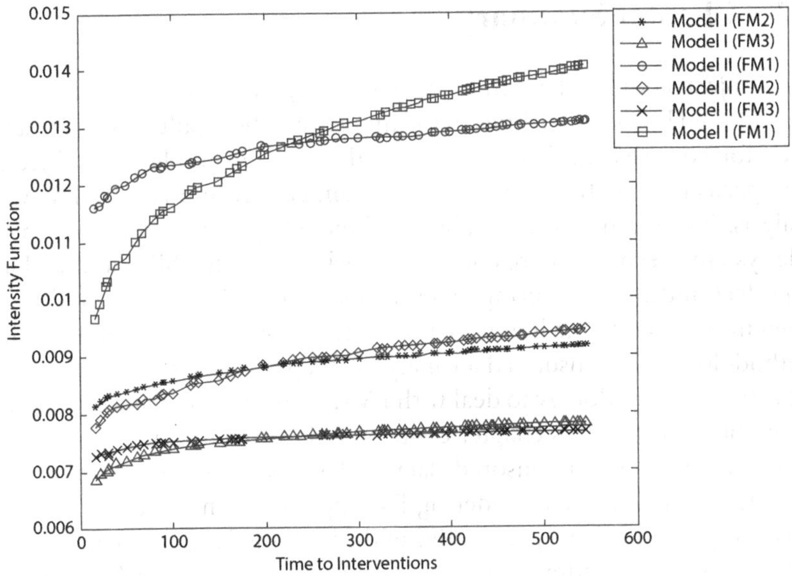

Figure 4.9 Intensity function graphs of the system considering dominant FMs.

actions will have a high possibility of removing damages that occurred till that time.

(c) With the help of the obtained parameters, the intensity function graphs are plotted for the system considering all three dominant FMs and presented in Figure 4.9. The graphs indicate that due to FM1, the system is deteriorating more as compared with the other two FMs.

Similarly, the technique and the models can also be used for:

1. Type II censoring if the test is failure-terminated instead of time-terminated. Everything remains the same.
2. Multiply censoring if the test is not terminated (time or failure) and the units are either removed from the test or we want to carry out just FM-wise analysis.

4.7 Chapter Summary

The subject addressed in this chapter is largely inspired by situations encountered by MRO facilities in the industries where failure data are analyzed for complex mechanical systems that are repairable and subjected to imperfect preventive and corrective maintenance and require FM-wise analysis. The mathematical viability of modeling a situation for repairable systems becomes more complicated when both the SPM and CM are imperfect and data censoring analysis due to FMs is also required to be taken into account. The literature is reasonably oriented toward providing methodologies for censored data analysis of repairable systems but lacks in providing a methodology to deal with FM-based censored data of complex repairable systems. This chapter contributes in providing a technique and virtual age models for censored data analysis of repairable systems with imperfect maintenance considering FM approaches in right and multiply censoring. An industrial perspective of the presented technique in the field of FMEA and RUL estimation and the conversion of the models to existing ones is also highlighted.

❖ Exercise Problem

Data on times to failures of air conditioning units of seven (K = 7) aircrafts are presented in Table 4.12. The purpose of the study is to conduct a reliability study with the help of three identified FMs and their failure times. Information of this type is useful in the analysis of many aspects of reliability, maintenance, and so forth. Since the aircraft is out of service if the air conditioning unit is inoperable, an effective maintenance policy is essential.

The test was time terminated at 2500 h. The aero engines are subjected to OEM-specified periodic PMs at fixed intervals of 100 operating hours (i.e., at SPM_1 = 100, SPM_2 = 200, SPM_3 = 300....).

(a) Considering model I and model II estimate the reliability parameters for each FM.

(b) Compare the results and examine the results which FM is causing the system deterioration.

Table 4.12 Time to failures of air conditioning unit.

System-wise failure times (in operating hours)	
System 1	194, 15, 41, 29, 33, 181
System 2	413, 427, 485, 522, 622, 687, 696, 865, 1312, 1496, 1532, 1733, 1851, 1885, 1916, 1934, 1952, 2019, 2076, 2138, 2145, 2167, 2201
System 3	90, 100, 160, 346, 407, 456, 470, 494, 550, 570, 649, 733, 777, 836, 865, 983, 1008, 1164, 1474, 1550, 1576, 1620, 1643, 1705, 1835, 2043, 2113, 2214, 2422
System 4	74, 131, 179, 208, 710, 722, 792, 813, 842, 1228, 1287, 1314, 1467, 1493, 1819
System 5	55, 320, 56, 104, 220, 239, 47, 246, 176, 182, 33, 15, 104, 35
System 6	23, 261, 87, 7, 120, 14, 62, 47, 225, 71, 246, 21, 42, 20, 5, 12, 120, 11, 3, 14, 71, 11, 14, 11, 16, 90, 1, 16
System 7	97, 51, 11, 4, 141, 18, 142, 68, 77, 80, 1, 16, 106, 206, 82, 54, 31, 216, 46, 111, 39, 63, 18, 191, 18, 163, 24
Failure mode-wise segregated data sets (in operating hours)	
FM1	194, 15, 41, 413, 427, 485, 522, 622, 687, 696, 865, 1312, 1496, 1532, 1474, 1550, 1576, 1620, 1643, 1705, 1835, 2043, 2113, 2214, 2422, 23, 261, 87, 7, 120, 14, 62, 47, 225, 71, 246, 21
FM2	1733, 1851, 1885, 1916, 1934, 1952, 2019, 2076, 90, 100, 160, 346, 407, 456, 470, 494, 550, 570, 710, 722, 792, 813, 842, 1228, 55, 320, 56, 104, 220, 97, 51, 11, 4, 141, 18, 142, 68, 77, 80, 1, 16, 106
FM3	2138, 2145, 2167, 2201, 29, 33, 181, 649, 733, 777, 836, 865, 983, 1008, 1164, 74, 131, 179, 208, 1287, 1314, 1467, 1493, 1819, 239, 47, 246, 176, 182, 33, 15, 104, 35, 42, 20, 5, 12, 120, 11, 3, 14, 71, 11, 14, 11, 16, 90, 1, 16, 206, 82, 54, 31, 216, 46, 111, 39, 63, 18, 191, 18, 163, 24

References

Bhattacharjee, P., Dey, V., Mandal, U.K., Risk assessment by Failure Mode and Effects Analysis (FMEA) using an interval number based logistic regression model. *Saf. Sci.*, 132, 104967, 2020.

Bichon, B.J., McFarland, J.M., Mahadevan, S., Efficient surrogate models for reliability analysis of systems with multiple failure modes. *Reliab. Eng. Syst. Saf.*, 96, 10, 1386–95, 2011.

Certa, A., Enea, M., Galante, G.M., La Fata, C.M., ELECTRE TRI-based approach to the failure modes classification on the basis of risk parameters: An alternative to the risk priority number. *Comput. Ind. Eng.*, 108, 100–110, 2017.

Crow, L.H., Reliability analysis for complex, repairable systems, ARMY MATERIEL SYSTEMS ANALYSIS ACTIVITYABERDEEN PROVING GROUND MD, *Reliability and Biometry*, 13, 6, 379–410, 1975.

Ebeling, C.E., *An Introduction to Reliability and Maintainability Engineering*, Tata McGraw-Hill Education, New York, United States, 2004.

Kijima, M., Some results for repairable systems with general repair. *J. Appl. Probab.*, 26, 1, 89–102, 1989.

Kim, K.O. and Zuo, M.J., General model for the risk priority number in failure mode and effects analysis. *Reliab. Eng. Syst. Saf.*, 169, 321–29, 2018.

Liu, H.C., Liu, L., Liu, N., Risk evaluation approaches in failure mode and effects analysis: A literature review. *Expert Syst. Appl.*, 40, 2, 828–38, 2013.

Meyer, R. and Krueger, D., *Minitab guide to statistics*, Prentice Hall PTR, Hoboken, New Jersey, U.S., 2001.

Nasr, A., Gasmi, S., Sayadi, M., Estimation of the parameters for a complex repairable system with preventive and corrective maintenance, in: *International Conference On Electrical Engineering and Software Applications (ICEESA)*, pp. 1–6, 2013.

Sharma, G. and Rai, R.N., A modified FMEA model with Repair Effectiveness Index (REI) using fuzzy AHP for life support systems of space stations, in: *International Conference on Applied Human Factors and Ergonomics*, Springer, pp. 77–87, 2018a.

Sharma, G. and Rai, R.N., Failure modes based censored data analysis for repairable systems and its industrial perspective. *Comput. Ind. Eng.*, 158, 107439, 2021b.

Sharma, G. and Rai, R.N., 9 - Modified failure modes and effects analysis model for critical and complex repairable Systems, in: *Safety and Reliability Modeling and Its Applications*, H. Pham and M. Ram (Eds.), pp. 245–60, Elsevier, Amsterdam, Netherlands, 2021e, https://doi.org/10.1016/B978-0-12-823323-8.00016-7.

Vassiliou, P., *Life data analysis reference, Weibull++ Version 6*, Reliasoft Corporation, Tucson, 2000.

Wang, H., Zhang, Y.M., Yang, Z., A risk evaluation method to prioritize failure modes based on failure data and a combination of fuzzy sets theory and grey theory. *Eng. Appl. Artif. Intell.*, 82, 216–25, 2019.

Wu, S., Warranty data analysis: A review. *Qual. Reliab. Eng. Int.*, 28, 8, 795–805, December, 2012. https://doi.org/10.1002/qre.1282.

Xiao, N., Huang, H.Z., Li, Y., He, L., Jin, T., Multiple failure modes analysis and weighted risk priority number evaluation in FMEA. *Eng. Fail. Anal.*, 18, 4, 1162–70, 2011.

5

Methodology for Identifying HFRCs Considering Risk-Based Threshold on Intensity Function

5.1 Introduction

While dealing with repairable systems, it is very difficult for an industry to control failures, manage maintenance activities, and achieve optimum availability at an optimal cost (Angus, Yin, and Trivedi, 2012). The MRO facilities in such industries are generally entrusted with these tasks. To control the number of failures and achieve high availability at less cost, these facilities (MRO) identify certain components such as HFRCs based on reliability analysis conducted on these components. Thus, the aim of designating HFRCs is to carry out the analysis for further reliability improvement (Xu *et al.*, 2017) which could be helpful to accomplish continuous operational performance. This chapter provides a solution for MRO facilities to deal with HFRCs and critical repairable systems. A brief discussion on HFRCs and the need for HFRC identification are carried out in the subsequent sections.

5.1.1 High Failure Rate Components (HFRCs)

The reliability of repairable systems is determined by their operational potential to accomplish targeted operations. Proper defect reporting and its investigations and reliability analysis of components make the organizations competently sound. The facilities of MRO (Asadi, Yu, and Mott, 2019) play a major role in taking obligatory actions on the components subjected to frequent failures. The arrivals of the same component due to repeated failures and other interventions, as acknowledged through the defect reporting and investigation database, are recognized and designated as HFRCs (Rai and Bolia, 2014).

Garima Sharma and Rajiv Nandan Rai. Advanced Techniques for Maintenance Modeling and Reliability Analysis of Repairable Systems, (111–128) © 2023 Scrivener Publishing LLC

5.1.1.1 Need for Identifying HFRCs

At present, the classification of HFRCs is done based on the number of failures, intuition/perceptions, and experience and sent for reliability improvement as a part of corrective measures. The components, failing in fielded units, are sent to the respective depots or at a required level of maintenance channel for CM. Along with such components, depots also induct components that are being sent for defect investigations and overhaul. Thus, the repair/overhaul lines get heavily loaded with repair and overhaul tasks. In such conditions, the HFRCs become an extra burden on depots. If more components are identified as HFRCs, then timely induction and production of components get impacted badly. As a result of the technique used for the declaration of HFRCs, many undesirable components get declared as HFRCs, and eligible components get omitted from the list of HFRCs. Thus, recognition of the actual HFRCs is necessary since too many components classified as HFRCs will lead to unwanted queuing as well as acquire undesirable extra costs for an organization. On the contrary, very few HFRCs may cause operational losses.

Evidently, the aforementioned reasons necessitate the requirement of a scientific methodology to decide a threshold for HFRC declaration, which will really help out in tiding over the crisis of more precise shortlisting, reducing the extra burden on the repair/overhaul lines, and placing the eligible components for reliability improvement, leading to the evasion of superfluous expenses incurred in taking improvement measures for undesirable components. In this area, the literature provides various cost and maintenance models/policies for repairable systems but lacks in delivering a methodology for HFRC declaration. However, the policy proposed by Rai and Bolia (2014) provides an availability-based methodology to address this problem but can be extended to an advanced level by introducing a risk-based intensity factor to it. Moreover, performance parameters can be estimated using a more realistic approach of GRP which considers both CM and SPM as imperfect.

The identification of HFRCs is important, but preventing the system from being an HFRC is also a very essential task. In this chapter, we will talk about a methodology to identify HFRCs for repairable systems by determining a risk-based threshold on the intensity function considering both CM and SPM as imperfect.

5.2 Methodology for Risk-Based Threshold on Intensity Function for HFRC Designation

To start with, one has to estimate the performance parameters using (maybe) the NHPP, Kijima, or Nasr *et al.* model for the given data set and calculate the corresponding virtual age at each intervention point. For failure rate, MTBF, and availability against each intervention time, the following expressions can be used (Sharma and Rai, 2021d):

Intensity function:

$$u(V_i) = \frac{1}{\eta^{\beta}} \times \beta \times (V_i)^{\beta-1} \quad (i = 1, 2, \ldots n_l) \tag{5.1}$$

Instantaneous mean time between failures (MTBFs) (Ebeling, 2004; Vassiliou, 2000):

$$MTBF(V_i) = \frac{1}{u(V_i)} \tag{5.2}$$

Availability:

$$A(V_i) = \frac{MTBF(V_i)}{MTBF(V_i) + MTTR} \tag{5.3}$$

where MTTR is the mean time to repair.

5.2.1 Steps of the Methodology

➢ *Step 1*: Find A^* to declare HFRC

- A^* is the minimum availability of the system that an organization can bear. The value of A^* depends on the operation demand of the organization.

➢ *Step 2*: Find t^*

- Having obtained A^*, find the time at which availability becomes less than or equal to A^*. This time is nothing but t^* at which the component should be declared as HFRC. Hence, for the HFRC zone,

$$A(t^*) = \frac{MTBF(t^*)}{MTBF(t^*) + MTTR} \leq A^* \tag{5.4}$$

➢ *Step 3*: Find the $Risk_{accept}$ value:
Let us consider t_{OH} as the time to overhaul a system. Then

- System availability (in hours) at any time "t_i" = $A(t_i) \times t_{OH}$
- System hours lost due to failures at any time "t_i" will be = $[1 - A(t_i)] \times t_{OH}$

Risk is defined as (Marhavilas, Koulouriotis, and Gemeni, 2011):

$$Risk\ (t_i) = F(t_i) \times Hours\_Lost \tag{5.5}$$

$$Risk\ (t_i) = F(t_i) \times t_{OH} \times [1 - A(t_i)] \tag{5.6}$$

Let $Risk_{accept}$ be the risk corresponding to the minimum required availability A^*. Hence,

- With the help of the t^* and A^* values, find the $Risk_{accept}$ with the help of the equation given below:

$$Risk_{accept} = F(t^*) \times t_{OH} \times [1 - A^*] \tag{5.7}$$

➢ *Step 4*: Find the λ^* value
The intensity function corresponding to $Risk_{accept}$ will be designated as λ^* which is the risk-based threshold value on intensity function. Thus, the zone for HFRC will be defined as:

$$\lambda(t^*) = \frac{\beta}{\eta^\beta} \times t^{*\beta-1} \geq \lambda^* \tag{5.8}$$

By applying this methodology, a risk-based threshold value $Risk_{accept}$ can be estimated and the corresponding λ value can be designated as λ^* which is the risk-based threshold value on intensity function.

✓ Solved Example
There are two variants (series 1 and series 2) of the same aero engines (turbojet engine with twin-spool compressor). The aero engines are

identical and put back into operation after general repair. Furthermore, the engines undergo periodic overhauls; the original equipment manufacturer (OEM) specified the time between overhaul (TBO) being 550 h. Within an overhaul cycle, the engines are subjected to OEM-specified periodic SPM at fixed intervals of 50 h (i.e., at $SPM_1 = 50$, $SPM_2 = 100$,.... $SPM_{11} = 550$ h (overhaul) as shown in Figure 5.1. The OEM-specified TTL of the engine is 1800 h. For series 1 and 2, there are 15 and 19 aero engines being considered with a mean time to repair (MTTR) = 76 h. The time-terminated multiple system data for series 1 and 2 engines are described in Tables 5.1 and 5.2, respectively. The tables contain both failure times and SPM times.

Considering the data set:

(a) Estimate the model parameters β, η, q_{CM}, and q_{SPM} using the Nasr *et al.* model and obtain the virtual age at each intervention time.

(b) Let us consider an aircraft that flies 20 days in a month. It can complete at least two sorties (one turn around) in a day and the duration of one sortie is 1 h. Using the methodology for HFRC designation, estimate the threshold parameters, i.e., A^*, t^*, λ^* and $Risk_{accept}$, for the aircraft (i) considering the total number of aircraft available in the inventory of unit per day as constant with varying flying hours essential for an aircraft in a month, and (ii) considering flying hours essential for an aircraft in a month as constant with the varying total number of aircraft available in the inventory of unit per day. Assume that the flying hours essential for an aircraft are 400, 410, 420, and 430 and the total number of aircraft available in the inventory of unit per day are 11, 12, 13, and 14.

(c) Define the HFRC zone for series 1 and series 2 aero engines. Plot the variation of threshold parameters for both cases and discuss the results in detail.

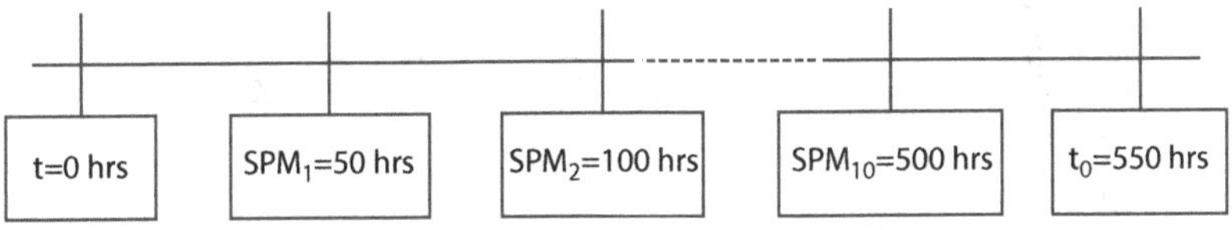

Figure 5.1 Classic PM policy.

Table 5.1 Time of Intervention (TOI) data since new for series 1 aero engines.

	System no.														
	1	2	3	4	5	6	7	8	9	10	11	12	13	14	15
TOI	50	50	49.02	50	50	50	50	19	50	50	50	14	50	50	14
	100	100	50	100	61	100	81	50	100	90	100	49	100	96	50
	150	132	100	114	100	150	100	100	142	100	106	50	150	100	100
	200	150	150	150	150	174	150	150	150	150	150	100	162	150	130
	230	200	200	200	200	200	184	200	200	200	200	150	169	164	131
	233	218	250	241	250	250	189	250	250	250	250	157	173	200	150
	250	250	300	250	300	300	200	280	300	300	300	200	200	226	172
	300	300	350	251	302	350	250	300	331	327	350	250	250	250	200
	350	307	400	277	305	400	300	321	350	350	361	300	300	300	250
	379	308	423	300	350	442	350	337	400	400	400	350	350	329	300

(Continued)

Table 5.1 Time of Intervention (TOI) data since new for series 1 aero engines. (*Continued*)

	System no.														
	1	**2**	**3**	**4**	**5**	**6**	**7**	**8**	**9**	**10**	**11**	**12**	**13**	**14**	**15**
TOI	400	350	450	350	400	450	388	350	441.58	415	442	356	365	350	340
	450	400	500	400	450	500	400	400	450	450	444	358	400	400	350
	464	450	536.52	428	500	536	450	450	500	497	450	400	450	450	400
	521	500	545.02	450	544.27	545	500	495	550	500	500	450	500	456	450
	550	550	550	500	550	550	545	500		550	504	465	550	500	500
				509			550	515			550	500		550	540
				550				550				550			550

Table 5.2 Time of Intervention (TOI) data since new for series 2 aero engines.

	System no.																		
	1	**2**	**3**	**4**	**5**	**6**	**7**	**8**	**9**	**10**	**11**	**12**	**13**	**14**	**15**	**16**	**17**	**18**	**19**
TOI	50	50	50	17	50	13	50	50	50	50	0.5	12	50	50	50	31	50	42	50
	100	100	100	50	100	15	100	100	100	100	50	50	100	100	100	50	100	47	100
	144	150	150	100	150	50	150	125	150	150	100	100	150	150	150	100	150	50	150
	150	200	200	150	200	100	200	150	200	198	150	150	200	200	200	150	166	100	179
	177	211	226	200	250	150	206	200	250	200	200	200	250	206	250	200	167	150	200
	200	250	229	250	273	200	250	250	300	250	250	250	300	250	300	150	200	200	250
	250	300	250	300	300	250	300	300	321	300	300	300	350	300	350	269	250	250	267
	300	350	261	350	350	300	350	350	350	304	350	336	400	350	366	300	300	300	300
	350	354	300	400	400	350	358	400	400	350	383	350	408	400	400	350	350	350	350
	361	400	350	425	401	400	400	450	450	400	400	400	450	450	450	400	390	387	400

(Coninued)

Table 5.2 Time of Intervention (TOI) data since new for series 2 aero engines. (*Continued*)

	System no.																		
	1	**2**	**3**	**4**	**5**	**6**	**7**	**8**	**9**	**10**	**11**	**12**	**13**	**14**	**15**	**16**	**17**	**18**	**19**
TOI	400	450	400	442	450	450	410	482	500	450	449	443	482	500	475	450	400	400	450
	450	500	450	450	500	500	450	500	544	469	450	450	484	549	500	500	450	450	500
	473	504	475	500	548	508	500	540	546	500	500	500	500	550	549	53	500	500	520
	500	550	488	550	550	550	515	550	550	550	550	501	548		550	550	534	548	543.5
	550		500				550					550	550				550	550	550
			550																

Solution:

(a) The parameters, estimated by maximizing the log-likelihood function of the Nasr *et al.* model (as given in Eq. (2.38)), are presented in Table 5.3. The corresponding virtual age at each intervention point is estimated using Eq. (2.35).

(b) Let us consider the flying hours essential for an aircraft, in a month to be X. Hence, hours of flying per day $Y = \frac{X}{20}$ by considering 20 days of flying in a month. Now let us consider that an aircraft can complete at least two sorties (one turn around) in a day and the duration of one sortie is 1 h. Thus, the required number of aircraft per day comes out to be $\frac{X}{40}$. If Z is the total number of aircraft available in the inventory of unit per day, then the minimum required availability of aircraft A^* will be equal to $\frac{X}{40Z}$. This value of A^* can be considered to be a measure of the minimum required availability.

Now, the variation in A^*, t^*, λ^*, and $Risk_{accept}$ is observed for two cases: (i) considering Z constant with varying X and (ii) considering X constant with varying Z. The threshold parameters for each case are appended in Tables 5.4–5.7.

Table 5.3 Estimated parameters.

Parameters	η	β	q_{SPM}	q_{CM}
Series 1	188.90	2.51	0.5	0.13
Series 2	274.16	3.73	0.12	0.44

Table 5.4 Threshold parameters considering constant Z in series 1.

X	Z	A^*	t^*	λ^*	$Risk_{accept}$
400	11	0.91	241.870	0.001301	4.53539
410	11	0.931	197.376	0.000975	2.00789
420	11	0.954	157	0.000631	0.71518
430	11	0.977	106	0.000306	0.10622

Table 5.5 Threshold parameters considering constant X in series 1.

X	Z	A^*	t^*	λ^*	$Risk_{accept}$
400	11	0.91	241.870	0.001301	4.53539
400	12	0.833	342.064	0.002638	24.7909
400	13	0.769	438.130	0.003952	58.4633
400	14	0.714	549.006	0.005270	99.8071

Table 5.6 Threshold parameters considering constant Z in series 2.

X	Z	A^*	t^*	λ^*	$Risk_{accept}$
400	11	0.91	353.190	0.001301	18.9987
410	11	0.931	316.400	0.000975	9.97021
420	11	0.954	261	0.000622	3.78302
430	11	0.977	200	0.000308	0.84010

Table 5.7 Threshold parameters considering constant X in series 2.

X	Z	A^*	t^*	λ^*	$Risk_{accept}$
400	11	0.91	353.190	0.001301	18.9987
400	12	0.833	438.320	0.002639	64.2097
400	13	0.769	502.379	0.003952	113.627
400	14	0.714	546	0.005256	151.327

(c) Let us consider a case where $X = 400$, $Z = 11$. The threshold parameters obtained in this case are given in Table 5.4 (first row values). The HFRC zone for this case is presented in Figure 5.2. The area covered with a solid black line is the safe zone for the considered case.

Similarly, the graphical representations of the threshold parameters considering each case presented in Tables 5.4–5.7 are shown in Figures 5.3–5.6.

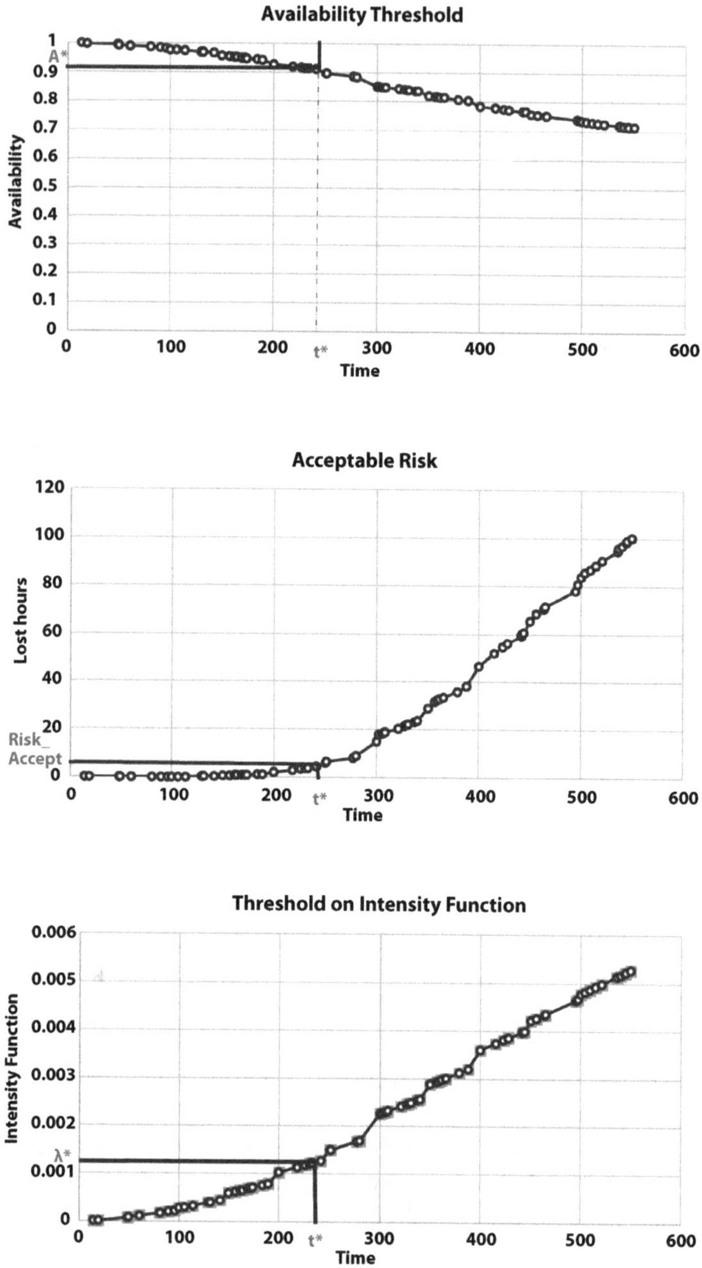

Figure 5.2 Representation of threshold parameters.

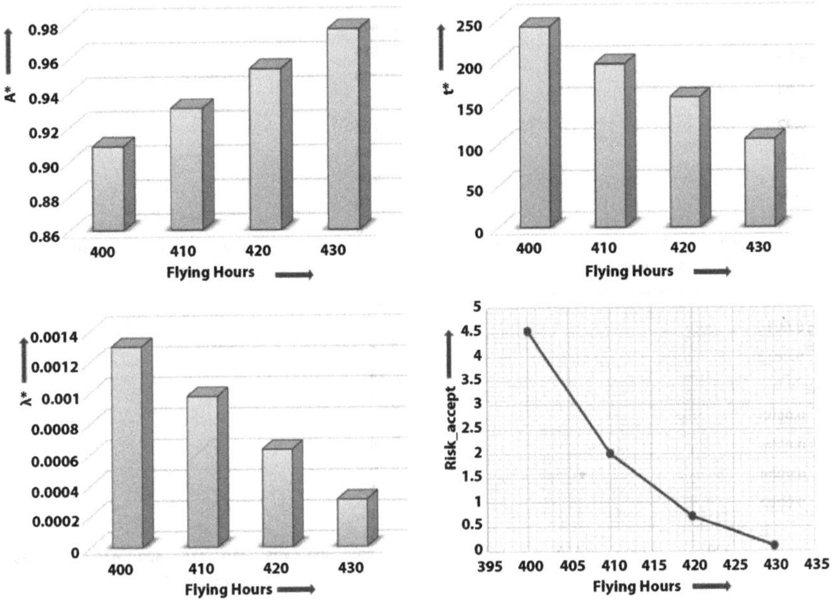

Figure 5.3 Variation threshold parameters keeping Z constant in series 1.

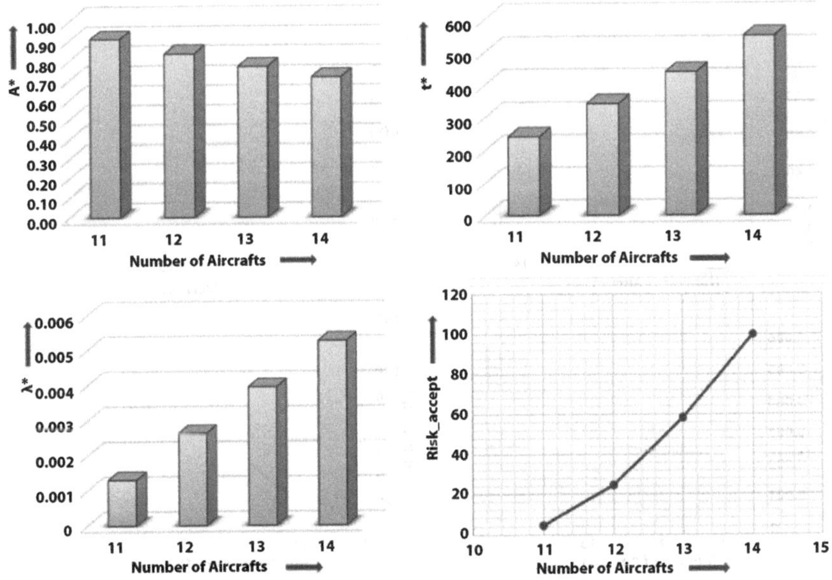

Figure 5.4 Variation in threshold parameters keeping X constant in series 1.

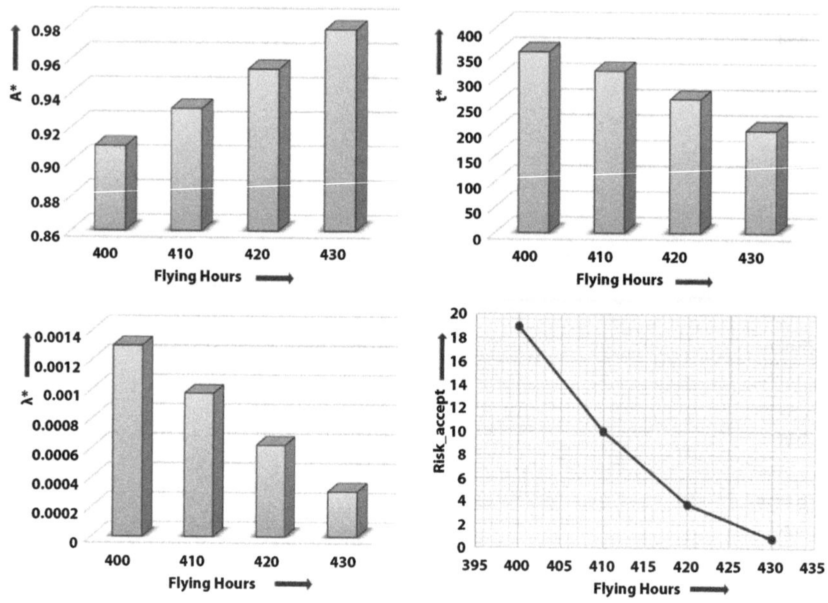

Figure 5.5 Variation in threshold parameters keeping *Z* constant in series 2.

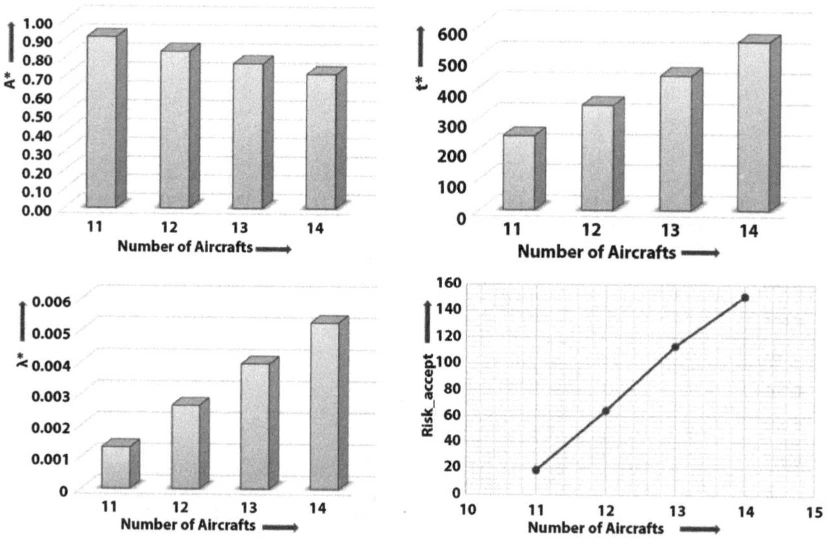

Figure 5.6 Variation in threshold parameters keeping *X* constant in series 2.

It can be observed from the above figures that, for both series 1 and series 2 aero engines, when we consider case (i), t^*, λ^*, and $Risk_{accept}$ decrease as A^* increases with X, whereas in case (ii), as A^* decreases with Z, hence, t^*, λ^*, and $Risk_{accept}$ increase.

If the aero engines are designated as HFRCs before they reach the estimated risk limits, the organization will fail in generating the required hours which will lead to the unavailability of aircrafts during missions. On the contrary, the designation of HFRCs beyond their estimated risk may result in some hazardous conditions. However, both cases would lead to a queuing of engines at the MRO waiting for repair and defect investigations.

Thus, with the help of HFRC designation methodology, the likelihood of designating the deserving components as HFRC increases which could result in preventing the MRO facilities from undue burden of components for repair or overhaul and enhancing its capacity to meet the unnecessary additional tasks leading to increased expenditures. In a nutshell, the threshold on risk would help in designating correct components as HFRCs leading to improved availability of engines with no extra cost, and resource augmentation results in the provisioning of adequate military aircraft for mission sorties.

5.3 Chapter Summary

The methodology presented in this chapter is highly inspired by an industrial problem of HFRC designation. The HFRCs have a high probability of failure and may require more apportionment of time and cost to do the targeted repair. Hence, it is required to apply a scientific methodology to decide a threshold for HFRC designation.

In this chapter, a methodology is discussed to establish a risk-based threshold on the intensity function for HFRC designation for repairable systems. With the help of such methodology, industries can estimate the threshold parameters, i.e., A^*, t, λ^*, and $Risk_{accept}$, for repairable systems according to their operational demand. With this, the likelihood of designating the deserving components as HFRC increases, which could result in (i) preventing the MRO facilities from the undue burden of components for repair or overhaul and (ii) enhancing its capacity to meet the unnecessary additional tasks leading to increased expenditures.

There could be one more solution to prevent the system from being classified as HFRC, i.e., applying such a maintenance policy which could be amended after each service activity according to the failure profile of the system. The same concept is extensively discussed in the next chapter.

Table 5.8 Failure data (flight hours) for jet engine.

Failure times	**Service times (non failure)**	
684	350	1350
701	650	1450
770	750	1550
812	850	1550
821	850	1650
845	950	1750
	550	1850
	1050	1850
	1050	1950
	1150	2050
	1150	2050
	1250	2050
	1250	

❖ Exercise Problem

A jet aircraft engine consists of many parts and, though generally very reliable, can fail in many ways. One approach to analyze data on multiple failure modes is to analyze each mode separately. Table 5.8 includes data of a military aircraft at a particular base. Of these, six had been removed and repaired prior to a scheduled maintenance action because of the failure of a turbine airfoil. Data are flight hours until failure for the failed items and approximate flight hours for nonfailed items (maintenance time). Data reported for nonfailed items were given in interval groupings; the data shown in Table 5.8 are the midpoints of the reported intervals.

(a) Estimate the performance parameters using any imperfect maintenance model.
(b) Let us consider an aircraft flies 15 days in a month. It can complete at least two sorties (one turn around) in a day and the duration of one sortie is 1 h. Estimate the threshold parameters,

i.e., A^*, t^*, λ^*, and $Risk_{accept}$, for the aircraft at 1850 flying hours with a total number of aircraft available in the inventory as 14.

(c) Show the threshold parameters with the help of a graphical representation.

References

Angus, J.E., Yin, M.L., Trivedi, K., Optimal random age replacement for availability. *Int. J. Reliab. Qual. Saf. Eng.*, 19, 05, 1250021, 2012.

Asadi, H., Yu, D., Mott, J.H., Risk factors for musculoskeletal injuries in airline maintenance, repair & overhaul. *Int. J. Ind. Ergon.*, 70, 107–15, 2019.

Ebeling, C.E., *An introduction to reliability and maintainability engineering*, Tata McGraw-Hill Education, New York, United States, 2004.

Marhavilas, P.K., Koulouriotis, D., Gemeni, V., Risk analysis and assessment methodologies in the work sites: On a review, classification and comparative study of the scientific literature of the period 2000–2009. *J. Loss Prev. Process Ind.*, 24, 5, 477–523, 2011.

Rai, R.N. and Bolia, N., Availability based optimal maintenance policies in military aviation. *Int. J. Perform. Eng.*, 10, 6, 641–48, 2014.

Sharma, G. and Rai, R.N., Risk-based threshold on intensity function of repairable systems: A case study on aero engines. *Int. J. Reliab. Qual. Saf. Eng.*, 28, 03, 2150020, 2021d.

Vassiliou, P., *Life Data Analysis Reference, Weibull++ Version 6*, Reliasoft Corporation, Tucson, 2000.

Xu, J., Luo, N., Yang, Y., Han, P., Zhang, D., Reliability analysis for the hydraulic booster control surface of aircraft. *J. Aircr.*, 54, 2, 456–63, 2017. https://doi.org/10.2514/1.C033500.

6
Progressive Maintenance Policy

6.1 Introduction

Maintenance, repair, and overhaul facilities, dealing with complex and critical repairable systems, often observe that the degradation profile of a system varies after each maintenance activity, and therefore, the typical maintenance policies (Zheng *et al.*, 2016; Lu, Cui, and Han, 2015), which consider SPM, age-based maintenance, and overhauls to be performed after a certain time interval, could be ineffective to estimate the real condition of the system. Due to the degradation profile of the system, the actual operational life of the equipment becomes much greater than the stipulated technical life. To further amplify this statement, let us consider a case of aero engines the details of which are presented in Section 5.2 (Solved Example). When the engine is operated till its first overhaul life, i.e., 550 h, because of wear out and degradation, its MRL would be much lesser than 550 h. Similarly, when the engine reaches its TTL of 1800 h, its MRL is much lesser than 1800 h. This indicates that, in reality, the engines having a stipulated life of 1800 h are being operated forcibly till a life greater than 1800 h. Therefore, traditional/classic maintenance lacks in capturing the wear-out phenomenon of the system which generally varies with time. As a consequence, the failure frequency increases, and the system gets identified as HFRC by the organization. The situation becomes worse when an organization plans for a life extension of the equipment based on the stipulated life. Since the traditional/classic maintenance policies do not consider MRL or the risk of failures of the system, the provided/estimated life extension may not be realistic and the system might face severe consequences in terms of major catastrophic failures.

Since the occurrence of such catastrophic failures in critical and complex repairable systems can end up in a very hazardous condition, such systems need continuous attention on their degradation/failure profiles and require a timely updated maintenance policy. Such policy also prevents the system from being in the category of HFRCs and increases the availability of the system.

Garima Sharma and Rajiv Nandan Rai. Advanced Techniques for Maintenance Modeling and Reliability Analysis of Repairable Systems, (129–140)

Concentrating on the above notion, a maintenance policy called progressive maintenance policy (PMP) is presented in this chapter considering MRS with both SPM and CM as imperfect in the KI model.

6.2 Progressive Maintenance Policy (PMP)

Definition: The system's hard life, SPM time, and overhaul period should be revised after each PM activity according to its failure profile and MRL in such a way that the risk of failure is not increased. Hard life refers here to the age at which the age-based SPM is to be done (Sharma and Rai, 2021c).

As defined above, unlike the other classic maintenance policies, the PMP constantly observes the failure profile and accordingly reviews SPM time and hard life of the system without increasing the risk of failure and also adjusts the time to overhaul based on MRL after each PM activity.

6.3 PMP Methodology

The following six steps describe the PMP methodology:

6.3.1 Steps of the PMP Methodology

➢ *STEP 1:* Estimation of hard life

Estimation of the hard life of the system could be done in different ways. For instance, let us consider the hard life (U. D. Kumar, Crocker, and Knezevic, 1999) $T_h = \left(\frac{MTBF_{at\ overhaul\ time}}{2} \right)$ hours.

As an example, for $i = 1$ (i.e., very first SPM), $t_{SPM_i} = 50$ and $t_{O_{R0}} = 550$ h (considering the case as given in Section 5.2 (Solved Example)). The iterative process starts from STEP 2 to obtain optimal overhaul time.

➢ *STEP 2:* Estimation of the ith SPM time (t_{SPM_i}) after each $(i - 1)^{th}$ SPM intervention

According to the definition of PMP, the subsequent SPM has to be planned in such a way that the risk of failure does not increase. For this purpose, the conditional reliability concept is used. If the conditional reliability (probability of survival till the ith SPM, if the system has already survived till $(i - 1)^{th}$ SPM) at i^{th} SPM is equal to the reliability at $(i - 1)^{th}$ SPM, then we can conclude that the risk of failure at i^{th} SPM is nearly the same as it was at $(i - 1)^{th}$ SPM. Let us denote reliability at $t_{SPM_{i-1}}$ as $R(t_{SPM_{i-1}})$. Hence, the following condition shown in Eq. (6.1) needs to be satisfied:

$$\frac{R(t_{SPM_i})}{R(t_{SPM_{i-1}})} = R(t_{SPM_{i-1}}) \tag{6.1}$$

$$R(t_{SPM_i}) = R(t_{SPM_{1-1}})^2 \tag{6.2}$$

Having obtained the reliability at i^{th} SPM, one can interpolate the corresponding V_i and hence obtain the revised time of ith SPM.

➢ *STEP 3:* Estimation of MRL after each ith SPM intervention
Having estimated the time of ith SPM, the next step is to estimate the MRL based on the failure profile of the system. This provides an idea about the next potential working hours of the system based on the failure history.

The mean residual life at any virtual time V_i is given in Eq. (6.3):

$$MRL(V_i) = \frac{\int_{V_i}^{\infty} R(V)dV}{R(V_i)} \tag{6.3}$$

Where V_i is given in Eq. (2.35).

Hence, for one overhaul cycle, *MRL* after i^{th} SPM can be expressed as:

$$MRL_i = \frac{\int_{V_{SPM_i}}^{V_{OR_{i-1}}} R(V)dV}{R(V_{SPM_i})} \tag{6.4}$$

➢ *STEP 4:* Review the overhaul time after each ith SPM intervention
The accumulated time of ith SPM and MRL, after the ith SPM, is the updated overhaul time of the system in terms of virtual age. This can be expressed as: $V_{O_{Ri}} = (MRL_i + V_{SPM_i})$ hours. The corresponding real time will be the revised overhaul time.

➢ *STEP 5:* Review the hard life time after each ith SPM intervention
As already described in STEP 1, the revised hard life after the ith SPM, i.e., $T_{h_{Ri}}$, comes out to be $(MTBF(V_{O_{Ri}})/2)$ hours. Hence, at this time, an age-based SPM is to be done.

➢ *STEP 6:* Iterative cycle termination
After each SPM, the policy should be revised until the $V_{O_{Ri}}$ value surpasses the $V_{O_{Ri-1}}$ value.

NOTE: At each step, the calculations should be carried out considering the virtual age, and thus, the corresponding real age should be estimated accordingly.

6.3.2 Motivation and Superiority of the Six-Step Maintenance Policy

Mechanical systems components are more prone to wear out as they age (i.e., their times to failure are age-related). The failure of one component may have a cascading effect; hence, at times, replacing the components before they fail could prove to be cost-effective, but the removal of items before they fail results in loss of useful life. It is also imperative to understand that maintenance cannibalization of components takes place. Cannibalization is a process in which failed components may be replaced with parts that have been recovered from previous repair. Thus, these components will have different time-to-failure distribution and wear-out characteristics. In addition, the safety critical items are mainly resorted to age-based preventive replacements as they are hard-lived. Hence, they are removed from the service as soon as the age of the part reaches the specified hard life. For these reasons, there is a requirement to adjust the initial maintenance after every maintenance event.

In PMP, the maintenance schedule is reviewed after every SPM activity without increasing the risk of failure. The hard life and next SPM activity are revised by adjusting the maintenance interval time. Thus, with the help of PMP, better availability with lesser risk could be achieved especially for critical and complex repairable systems.

✓ Solved Example
Using the data sets (series 1 and series 2) and the performance parameters as described in Section 5.2 (Solved Example) and Table 5.3, respectively:

(a) Develop a revised maintenance policy by applying the PMP concept for series 1 and series 2 aero engines and explain the results in detail.
(b) Perform a comparative study between the present and the PMP for both cases considering MTBF, availability, and total cost.

Solution:
Based on the first overhaul failure data, the recommended maintenance plans for the next overhaul cycle using the PMP methodology for both cases (series 1 and series 2) are presented in Tables 6.1 and 6.2, respectively.

Table 6.1 Progressive maintenance plan for series 1 aero engines.

Sl no. (i^{th} PM)	**Original SPM schedule (SPM_i) (h)**	**SPM time after applying PMP(t_{SPM_i}) (h)**	**Mean residual life (MRL_i) (h)**	**Revised hard life (T_{hRi}) (h))**	**Revised time to overhaul (t_{ORi}) (h)**
1	50	50	460.28	102.12	504.52
2	100	73.85	417.21	110.17	481.20
3	150	94.80	391.51	117.18	449.47
4	200	118.00	348.77	123.06	445.64
5	250	151.47	310.74	127.02	436.45
6	300	193.33	287.73	128.08	433.00
7	350	–	–	–	–
8	400	–	–	–	–
9	450	–	–	–	–
10	500	–	–	–	–
11	550	–	–	–	–

Table 6.2 Progressive maintenance plan for series 2 aero engines.

Sl no. (i^{th} PM)	Original SPM schedule (SPM_i) (h)	SPM time after applying PMP(t_{SPM_i})(h)	Mean residual life (MRL_i) (h)	Revised hard life ($T_{h_{Ri}}$)(h)	Revised time to overhaul ($t_{O_{Ri}}$)(h)
1	50	50	469.04	108.50	525.50
2	100	102.50	433.79	122.30	507.31
3	150	117.50	404.56	135.07	484.25
4	200	132.12	371.63	146.20	473.20
5	250	143.50	356.50	155.00	465.63
6	300	167.00	320.18	159.16	462.11
7	350	–	–	–	–
8	400	–	–	–	–
9	450	–	–	–	–
10	500	–	–	–	–
11	550	–	–	–	–

Note: The corresponding virtual age for each intervention is estimated using Eq. (2.35).

One hand calculation considering series 2 aero engines

For the first cycle:

- $t_{SPM_1} = 50\,\text{h}$ (the corresponding virtual age is 21.04 h) and overhaul at 550 h (the corresponding virtual age is 195 h):
- $$MRL_1 = \frac{\int_{21.04}^{195} e^{-(8\times10^{-10}\times V^{3.73})}dV}{R(21.04)} = 163.50$$
  $= 469.04\,\text{h}$ (in real time scale)
- $V_{O_1} = (163.50 + 21.04) = 184.50$

Thus, $t_{O_{R1}} = 525.50\,\text{h}$ (in real time scale)

- $$T_{h_{R1}} = (MTBF(V_{O_1})/2) = \frac{217.00}{2} = 108.5\,\text{h}$$

For the second cycle:

- $R(t_{SPM_2}) = R(t_{SPM_{1-1}})^2 = (0.9991)^2 = 0.9982;$
  $R(102.50) = 0.9982$
- Thus $t_{SPM_2} = 102.50\,\text{h}$ and the corresponding virtual age is 28.14 h.
- $$MRL_2 = \frac{\int_{28.14}^{184.50} e^{-(8\times10^{-10}\times V^{3.73})}dV}{R(28.14)} = 148.31$$
  $= 433.79\,\text{h}$ (in real time scale)
- $V_{O_2} = (148.31 + 28.14) = 176.50$

Thus, $t_{O_{R1}} = 507.31\,\text{h}$ (in real time scale)

- $$T_{h_{R1}} = (MTBF(V_{O_2})/2) = \frac{244.60}{2} = 122.30\,\text{h}$$

Continue the cycle until the $V_{O_{Ri}}$ value surpasses the $V_{O_{R_{i-1}}}$.

In the classic/traditional maintenance policy applied for aero engines in aviation industries as shown in Figure 5.1, the number of SPM is 10, i.e., every 50 till 500 h, and the overhaul is completed at 550 h. It can be observed from

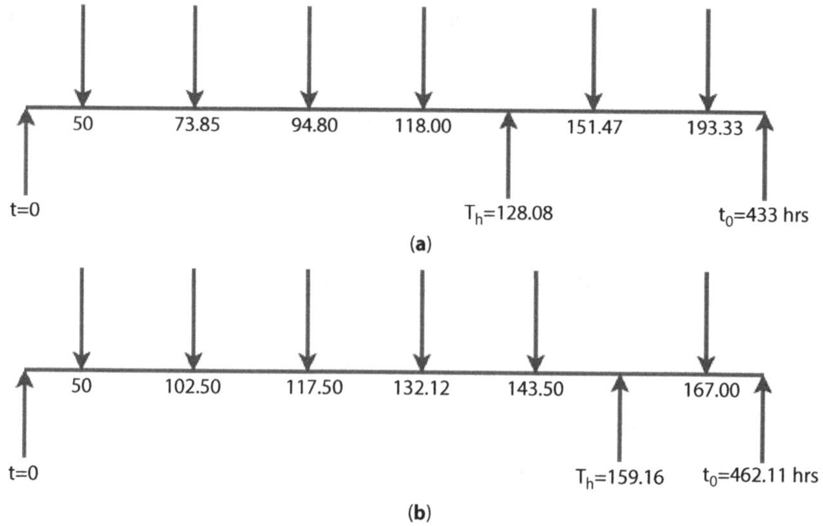

Figure 6.1 PMP for (a) series 1 aero engines and (b) series 2 aero engines.

Tables 6.1 and 6.2 that the number of SPMs is rescheduled according to the failure profile of the respective cases. An additional age-based PM (maintenance at hard life) at 128.08 h with a reviewed overhaul time of 433 h for series 1 aero engines is obtained, whereas for series 2 engines, one age-based PM is obtained at 159.16 h and the reviewed overhaul time period is 462.11 h (Figure 6.1). It can also be observed that the hard life gets extended by 34.651 (according to hand calculation) and 65.95 h for series 1 and series 2, respectively.

The results show that the time period between two SPM cycles is less initially for the PMP as compared with the classic/traditional SPM policy but gradually increases, and the total number of SPMs is reduced from 10 to 6 in both cases. In addition, the TBO cycle is also getting reduced in both cases.

(e) Comparative Study Between Classic Maintenance Policy and PMP

The comparison matrix is shown in Tables 6.3 and 6.4. The availability trend for both cases is also presented in Figures 6.2 and 6.3, respectively. Increased availability is found at each revised SPM time for both cases as compared with the classic/traditional maintenance policy.

Note: For failure rate, MTBF, and availability against each intervention time, the expressions given in Eqs. (5.1)–(5.3) are used.

With the help of the PMP, significant improvements in MTBF, availability, and maintenance cost are witnessed. For series 1 aero engines, an

Table 6.3 Comparative results for series 1 aero engines.

Performance parameters at the end of the overhaul cycle			
Maintenance policies	**MTBF (t_o)**	**A (t_o)**	**Cost (if x is the cost of one SPM)**
Classic maintenance policy	189.37	0.7136	10x
Progressive maintenance policy	256.17	0.7711	6x
Percent improvement	35.27	8.05%	40%

Table 6.4 Comparative results for series 2 aero engines.

Performance parameters at the end of the overhaul cycle			
Maintenance policies	**MTBF (t_o)**	**A (t_o)**	**Cost (if x is the cost of one SPM)**
Classic maintenance policy	186.42	0.7103	10x
Progressive maintenance policy	318.32	0.8066	6x
Percent improvement	70.75	13.55%	40%

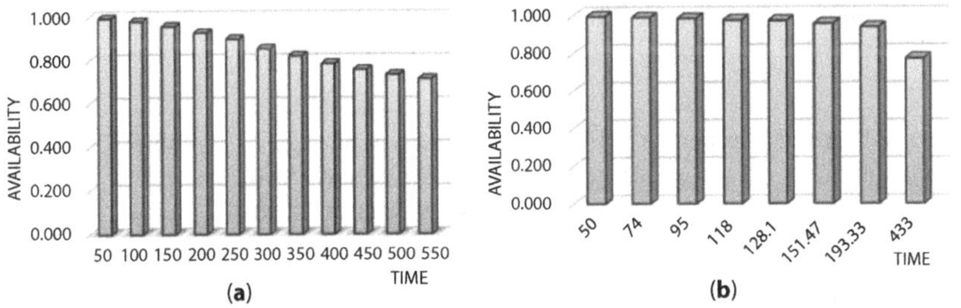

Figure 6.2 Availability trend for series 1 aero engines: (a) classic PM policy and (b) PMP.

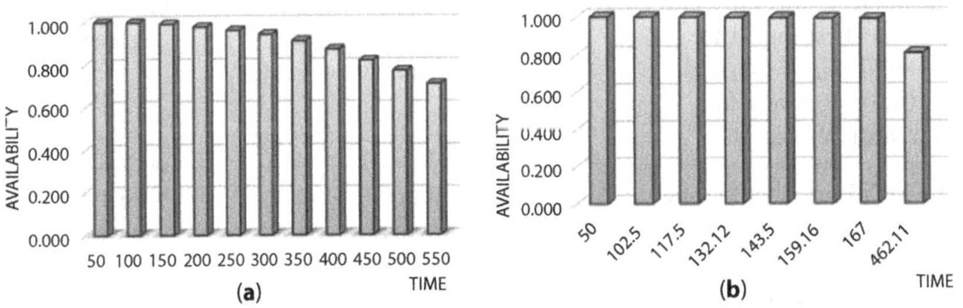

Figure 6.3 Availability trend for series 2 aero engines: (a) classic PM policy and (b) PMP.

improvement of 35.27% in MTBF and 8.05% in availability is noticed. Similarly, for series 2 aero engines, an improvement of 70.75% in MTBF and 13.55% in availability is observed. There is a 40% improvement in maintenance cost (approximate estimation) that could be achieved through PMP for both the considered cases. The comparative study is also summarized in Table 6.5 for better understanding.

Table 6.5 Comparative study between classic and progressive maintenance policies.

Classic maintenance policy	PMP	Remarks
Considers SPM and overhauls to be done at a fixed time interval	Hard life, SPM scheduled time, and overhaul period of the system are revised after each service activity which can be done by adjusting the PM interval and mean residual life (MRL) such that the risk of failure is not increased.	Six-step maintenance as explained in Section 6.3.1
May fail to monitor the exact condition of the component	Continuously observes the failure profile of the system after each maintenance activity, hence can monitor the condition of the equipment more precisely.	Six-step maintenance as explained in Section 6.3.1
Total number of time-based SPMs carried out = 10	The number of SPMs is reduced from 10 to 6 for both series of aero engines.	Tables 6.1 and 6.2
No age-based PMs as per hard life	There is an addition of one age-based PM at 128.08 and 159.16 h for series 1 and series 2 aero engines, respectively.	Tables 6.1 and 6.2
The original TBO cycle of both series of aero engines is 550 h.	The revised TBO is 433 and 462.11 h for series 1 and 2, respectively.	Tables 6.1 and 6.2
The MTBF, availability, and the cost estimated are 189.370 h, 0.7136, and 10x, respectively, for series 1 aero engines.	An improvement of 35.27% in MTBF, 8.05% in availability, and 40% in maintenance cost is observed.	Table 6.3 and Figure 6.2
The MTBF, availability, and the cost estimated are 186.42 h, 0.7103, and 10x, respectively, for series 2 aero engines.	An improvement of 70.75% in MTBF, 13.55% in availability, and 40% in maintenance cost is observed.	Table 6.4 and Figure 6.3

In this chapter, the cases of aero engines from the aviation industry are considered, but a similar approach can be applied to other critical and complex repairable systems which undergo similar maintenance policies in their TTL.

6.4 Chapter Summary

In this chapter, a PMP, based on the risk and MRL considering MRS with both SPM and CM as imperfect in the KI model, is presented. According to the PMP, the hard life, SPM time, and overhaul time of the system are revised after each SPM activity which is done by adjusting the SPM interval in such a way that the risk of failure does not increase. A PMP could be more beneficial for industries that deal with complex and critical repairable systems such as automobile industries, nuclear power plants, and aerospace industries.

The PMP continuously gets amended according to the failure profile of the system and, hence, is more competent and flexible in its application than any other maintenance policies. The policy can be extended to any other repairable systems which follow the maintenance policy similar to the classic PM policy.

❖ Exercise Problem

Please refer to the Exercise Problem of Chapter 5 (Table 5.8). A jet aircraft engine consists of many parts and, though generally very reliable, can fail in many ways. One approach to analyze data on multiple failure modes is to analyze each mode separately. Table 5.8 includes data of a military aircraft at a particular base. Of these, six had been removed and repaired prior to a scheduled maintenance action because of failure of a turbine airfoil. Data are flight hours until failure for the failed items and approximate flight hours for nonfailed items (maintenance time). The data reported for nonfailed items were given in interval groupings; the data shown in Table 5.8 are the midpoints of the reported intervals. The PMP for the given data set is developed and the variation in the availability of the aircraft is analyzed.

References

Dinesh Kumar, U., Crocker, J., Knezevic, J., Evolutionary maintenance for aircraft engines, in: *Annual Reliability and Maintainability. Symposium. 1999 Proceedings (Cat. No. 99CH36283)*, IEEE, pp. 62–68, 1999.

Lu, Z., Cui, W., Han, X., Integrated production and preventive maintenance scheduling for a single machine with failure uncertainty. *Comput. Ind. Eng.*, 80, 236–44, 2015.

Sharma, G. and Rai, R.N., Progressive maintenance policy for multiple repairable systems with imperfect maintenance. *J. Qual. Maint. Eng.*, 28, 3, 668–85, 2021c.

Zheng, Z., Zhou, W., Zheng, Y., Wu, Y., Optimal maintenance policy for a system with preventive repair and two types of failures. *Comput. Ind. Eng.*, 98, 102–12, 2016.

7

Age-Based Maintenance Policies for Repairable Systems

7.1 Introduction

Reliability analysis of repairable systems is carried out by treating the system either as a single repairable system (SRS) or multiple repairable systems (MRS) (Garmabaki *et al.*, 2016). In SRS, the failure data of a particular system are considered for reliability analysis, whereas in MRS, the failure data of all identical systems working in the field are utilized for reliability analysis. Hence, MRS analysis portrays a larger picture of the system's condition compared with SRS. Both systems are represented in Figure 7.1.

This chapter provides optimal age-based overhaul policies using downtime optimization which assumes both SPM and CM as imperfect for RS. A spare parts estimation model for RS considering both the SPM and CM as imperfect is also discussed.

This chapter is structured into two parts: (1) two age-based overhaul policies are presented for RS using the KI model to obtain age-based optimal overhaul time when (a) only CM is imperfect (policy I) and (b) both SPM and CM are imperfect (policy II) and (2) a spare part estimation model for RS considering both the SPM and CM as imperfect for the KI model.

7.2 Age-Based Policies for Repairable Systems

In some cases, say, due to difficulties in determining costs or the desire to get maximum throughput or utilization of equipment, the overhaul policy required may be one that minimizes total downtime per unit time or, equivalently, maximizes availability (Sharma and Rai, 2021a). As the overhaul frequency increases, there is an increase in downtime but the consequence of this is the reduction of downtime due to less number of repair actions; hence, the best balance between them is required to be resorted

Garima Sharma and Rajiv Nandan Rai. Advanced Techniques for Maintenance Modeling and Reliability Analysis of Repairable Systems, (141–150) © 2023 Scrivener Publishing LLC

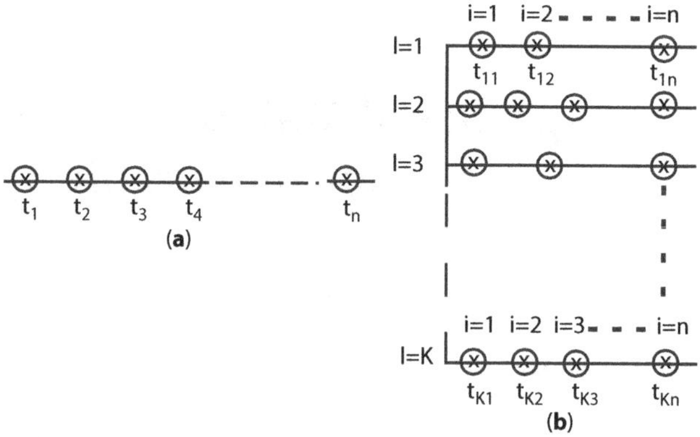

Figure 7.1 (a) Single repairable system and (b) multiple repairable systems.

to. Hence, this section presents two different age-based overhaul policies which determine the best age at which overhaul should occur to minimize total downtime per unit time. Both policies are based on the KI model with the downtime-based optimization technique. A spare parts estimation model is then presented for both the overhaul policies.

7.2.1 Age-Based Overhaul Period Estimation with Imperfect CM (Policy I)

The aim is to establish an optimal age for overhaul for the next overhaul cycle with the help of existing failure data of the previous overhaul cycle so that a high occurrence of failures may be prevented. Policy I is explained with the help of Figure 7.2. The methodology for downtime per unit time estimation is explained as follows.

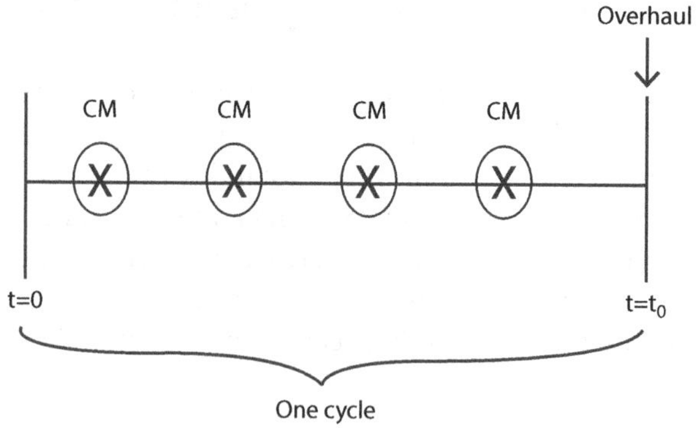

Figure 7.2 Age-based overhaul period estimation with imperfect CM (policy I).

Let D_R be the mean downtime required to perform a general repair, and D_O be the downtime required to perform an overhaul. The overhaul policy is to perform overhaul at age t_O, and general repairs occur as many times as required in the interval $(0,t_O)$.

The total downtime per unit time, for overhaul at age t_O, denoted by $D(t_O)$ (Jardine & Tsang, 2005) is:

$D(t_O)$ = Expected downtime due to repair + Expected downtime due to overhaulExpected Cycle length (7.1)

Expected downtime due to repair = Mean downtime due to repair cycle (D_R) × Probability of repair cycle $(1-R(t_O))$

Expected downtime due to overhaul = Downtime due to overhaul (D_O) × Probability of overhaul $R(t_O)$

Expected cycle length = Overhaul cycle length + Repair cycle length

$$\text{Overhaul cycle length} = (t_O + D_O)R(t_O)$$

$$\text{Repair cycle length} = [M(t_O) + D_R][1 - R(t_O)]$$

where

$$M(t_O) = \int_0^{t_O} t.f(t)dt \big/ 1 - R(t_O)$$

Therefore

$$D(t_O) = \frac{D_O R(t_O) + D_R[1 - R(t_O)]}{(t_O + D_O)R(t_O) + [M(t_O) + D_R][1 - R(t_O)]} \tag{7.2}$$

Using the above equation, the optimized minimum value of $D(t_O)$ can be obtained.

7.2.2 Age-Based Overhaul Time Estimation with Imperfect CM and SPM (Policy II)

This section describes the methodology of the age-based policy considering both CM and SPM in one overhaul cycle. Here, both CM and SPM are considered imperfect. It is considered that:

The operating units of the system are overhauled at every t_O, i.e., at every Pt_O, where t_O is the time to overhaul and P = 1,2,3, ...

If the item fails, while in operation at time $t_f < t_O$, i.e., in interval $[(P-1)t_O, (P-1)t_O + t_f]$, it is subjected to imperfect repair. In addition, the performed PM is scheduled at every 50 h (let us assume) which is also considered imperfect (refer to Figure 7.3).

The first step is to obtain the reliability estimation parameters, i.e., β, η, q_{CM}, and *qSPM* and corresponding virtual ages of the field failure data which can be estimated using Eq. (2.38).

Now the downtime equation can be written as:

$$D(t_O) = D_O + D_R E\left[N[0,t_O]\right] + \frac{t_O}{50} DSPM \tag{7.3}$$

$E\left[N[0,t_O]\right]$: Expected number of failures in interval $[0, t_O]$

$$E\left[N[0,t_O]\right] = \int_0^{t_O} u(t)\,dt;$$

where $u(t) = \frac{\beta}{\eta}\left(\frac{t}{\eta}\right)^{\beta-1}$ which is an intensity function of the power law model.

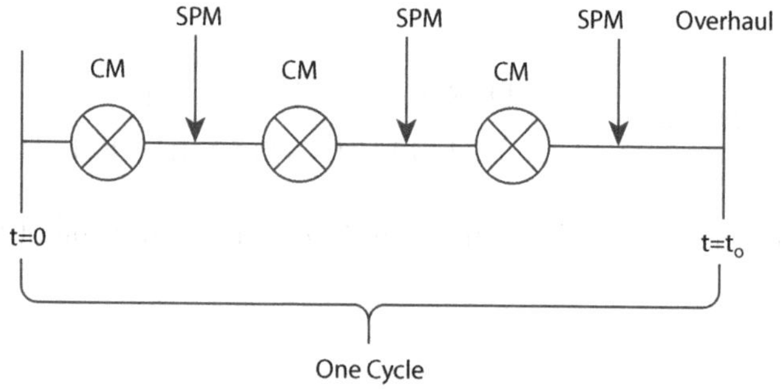

Figure 7.3 Age-based overhaul period estimation with imperfect CM and SPM (policy II).

The corresponding virtual age at t_O will be V_O. Thus,

$$E[N[0,V_O]] = \int_0^{V_O} u(v)dv = \frac{V_0^{\beta}}{\eta^{\beta}} \tag{7.4}$$

$$D(V_O) = D_O + D_R \times aV_0^{\beta} + \frac{V_O}{50} DSPM \tag{7.5}$$

where $a = \frac{1}{\eta^{\beta}}$;

Downtime per unit time can be expressed as:

$$\frac{D(V_O)}{V_O} = \frac{1}{V_O} D_O + D_R \times aV_0^{\beta} + \frac{V_O}{50} DSPM \tag{7.6}$$

By differentiating the above equation and equating it to zero, the optimal V_O value can be obtained as:

$$\frac{d}{dV_0}\left[\frac{D(V_O)}{V_O}\right] = \frac{d}{dV_0}\frac{1}{V_O} D_O + D_R \times aV_0^{\beta} + \frac{V_O}{50} DSPM = 0 \tag{7.7}$$

Hence, the optimal V_O can be estimated as:

$$V_0 = \left[\frac{D_O}{aD_R(\beta - 1)}\right]^{\frac{1}{\beta}} \tag{7.8}$$

Substituting the values of D_O, D_R, β, and a in Eq. (7.8), one can estimate the virtual optimal time for overhaul on the virtual scale and the corresponding overhaul time on the real-time scale.

7.2.3 Spare Parts Estimation

In this section, a methodology is presented on spare parts estimation for age-based overhaul policy considering both the SPM and CM as imperfect.

Spare parts estimation (Kececioglu, 2003) in the time interval $(t_2 - t_1)$, can be formulated as:

$$C_{SP}(t_2 - t_1) = C_T \times E[C_F(t_2 - t_1)] \tag{7.9}$$

where

C_{SP} is the total number of expected spare parts in interval $(t_2 - t_1)$

C_T is the total number of spare parts in use during the life period $(t_2 - t_1)$. It is determined either by monitoring such parts or by the theory of maintainability and the PM and CM policy and the schedule used.

$E[C_F(t_2 - t_1)]$ is the expected numbers of failure parts in interval $(t_2 - t_1)$

Thus:

$$C_{SP}(t_2 - t_1) = C_T \times \int_{t_1}^{t_2} u(t)\,dt \tag{7.10}$$

But in this case:

$$C_{SP}(t_O - 0) = C_T \times \int_{0}^{t_O} u(t)\,dt \tag{7.11}$$

and in terms of virtual age:

$$C_{SP}(V_O - 0) = C_T \times \int_{0}^{V_O} u(v)\,dv \tag{7.12}$$

Hence

$$C_{SP}(V_O - 0) = C_T \times aV_0{}^{\beta} \tag{7.13}$$

With the values of C_T, a, V_0, and β, one can estimate the number of spare parts for a particular overhaul time. In addition, the mean time between failures (MTBFs) and availability at each data point can be estimated using the following equations:

Mean time between failures (MTBFs)

$$MTBF(V_i) = \frac{1}{u(V_i)} \tag{7.14}$$

Availability

$$A(v_i) = \frac{MTBF(v_i)}{MTBF(v_i) + D_R} \tag{7.15}$$

✓ Solved Example

A turbojet aero engine (a repairable system) is subjected to general repairs on failing. The aero engines undergo periodic overhauls; the original equipment manufacturer (OEM) specified the time between overhaul (TBO) to be 550 h. The time to failures is appended in Table 7.1. Within an overhaul cycle, the aero engines are subjected to OEM-specified periodic time-based SPM at fixed intervals of 50 h (i.e., at $PM_1 = 50$, $PM_2 = 100$, $PM_3 = 150$.... $t_o = 550$ h). The existing PM policy for the aero engines is illustrated in Figure 7.4.

If the value of $D_O = 1900$ h, the mean time to repair $D_R = 528$ h, and the number of aero engines in use $C_T = 16$. Estimate:

(a) Reliability parameters for the given failure data.
(b) Age-based optimal overhaul time using policy I and policy II and the number of spare parts.
(c) Analyze the obtained results.

Table 7.1 Time to failure data.

Time to failures
21, 22, 23, 29, 34, 80, 102, 104, 106, 107, 137, 144, 155, 174, 174, 179, 197, 204, 206, 218, 231, 233, 241, 248, 263, 274, 279, 287, 288, 303, 304, 305, 309, 310, 316, 321, 326, 332, 345, 346, 371, 378, 391, 402, 408, 410, 410, 414, 416, 417, 423, 425, 428, 456, 474, 485, 487, 491, 495, 497, 502, 515, 516, 517, 522, 529, 530, 532, 535, 537, 550

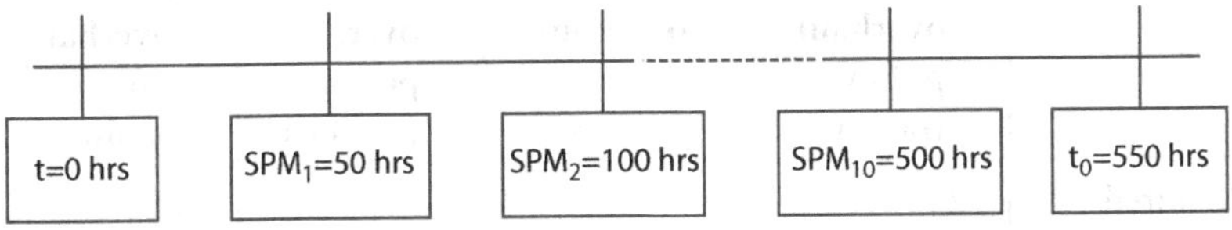

Figure 7.4 Present maintenance policy.

Solution:

(a) Reliability parameters are estimated by maximizing Eq. (2.38) and are tabulated in Table 7.2.

(b) The optimal overhaul time estimated for policies I and II as explained in Sections 7.2.1 and 7.2.2, respectively, are presented in Table 7.3. The estimated spare parts as explained in Section 7.2.3 for both policies are appended in Table 7.4.

(c) It can be observed from Tables 7.3 and 7.4 that the reviewed overhaul age with policy I is 416 h as compared with the existing overhaul time of 550 h. The availability improvement (using Eq. (7.15)) is estimated to be 9.87% and the number of spare parts that are required for the operation

Table 7.2 Estimated reliability parameters for policies I and II.

Types of models	$E[C_F(V_0 - 0)]$	β	η (h)	q_{CM}	q_{SPM}
Considering imperfect CM (Kijima I model) (policy I)	1.80	1.4	191.1173	0.7	–
Considering imperfect CM and PM (Nasr *et al.* model) (policy II)	2.89	2.25	169.534	0.6	0.4

Table 7.3 Estimated optimal overhaul time for policies I and II.

Policy	t_O (h)
Policy I	416
Policy II	485

Table 7.4 Estimated spare parts for policies I and II.

	Present overhaul policy (policy I)	Present overhaul policy (policy II)	Reviewed overhaul policy (policy I)	Reviewed overhaul policy (policy II)
Required spare parts	42	62	28	46

is 28. Hence, there is a reduction in the inventory of spare parts from 42 to 28. Thus, there is a significant savings of (42 – 28) = 14 aero engines resulting in a quantum savings in inventory cost.

It can be further noticed that the reviewed overhaul age with policy II is 485 h as compared with existing overhaul time of 550 h. The availability improvement is estimated to be 16.6% and the number of spare parts that are required to be stored is 46. Hence, there is a reduction in the inventory of spare parts from 62 to 46. Thus, there is a significant savings of (62 – 46) = 16 aero engines, once again resulting into a desirable reduction in the inventory cost.

Hence, it is worth implementing the age-based overhaul policies (policy I or II) in place of the existing overhaul policy. Both policies deliver not only a significant improvement in the availability of aero engines but also considerable savings in inventory cost. Comparing policies I and II, policy II is more practical as the consideration of both SPM and CM as being imperfect is taken into account in this case. Also, policy II provides better results in terms of availability improvement.

7.3 Chapter Summary

The method to obtain the optimal time for age-based policy in this chapter is largely inspired by the industries which deal with repairable systems and face problems in estimating an optimized overhaul time for deteriorating mechanical systems. The chapter discusses two age-based overhaul policies (policy I and policy II). Policy I considers only CM as imperfect to review the optimal overhaul time, whereas policy II considers both CM and SPM as imperfect. The chapter also presents a spare parts estimation model for the abovementioned overhaul policies (policy I and policy II) to evaluate the number of spares required in the inventory.

It is recommended to implement the age-based overhaul policies (policy I or II) in place of existing overhaul policy. Using such policies will provide not only a significant improvement in the availability of RS but also considerable savings in inventory cost. The policies can be applied to any complex MRS that undergoes general repair (CM), SPM, and overhaul during its life cycle.

Table 7.5 Time to failures of aero engine.

Time to failures
0.5, 12, 13, 15, 17, 31, 38, 39, 42, 47, 52, 67, 98, 101, 125, 133, 144, 166, 167, 177, 179, 189, 198, 206, 211, 212, 226, 264, 267, 269, 273, 293, 298, 321, 336, 344, 354, 361, 366, 383, 387, 390, 401, 408, 425, 443, 461, 475, 507, 520, 542, 544, 545, 547, 548, 549

❖ Exercise Problem

A turbojet aero engine (a repairable system) is subjected to general repairs on failing. The aero engines undergo periodic overhauls; the original equipment manufacturer (OEM) specified the time between overhaul (TBO) being 550 h. The time to failures is appended in Table 7.5. Within an overhaul cycle, the aero engines are subjected to OEM-specified periodic time-based SPM at fixed intervals of 50 h (i.e., at $PM_1 = 50$, $PM_2 = 100$, $PM_3 = 150$.... $t_o = 550$ h).

If the value of $D_O = 1900$ h, the mean time to repair $D_R = 528$ h, and the number of aero engines in use $C_T = 16$. Estimate:

(a) Estimate the reliability parameters considering CM and SPM as imperfect.
(b) Estimate the optimal overhaul time considering the present (as given in the question), policy I and policy II, and compare the results.
(c) Number of spare parts for the given data set to $t_o = 550$ h.

References

Garmabaki, A.H.S., Ahmadi, A., Mahmood, Y.A., Barabadi, A., Reliability modelling of multiple repairable units. *Qual. Reliab. Eng. Int.*, 32, 7, 2329–43, 2016.

Jardine, A.K.S. and Tsang, A.H.C., *Maintenance, replacement, and reliability: Theory and applications*, CRC Press, Boca Raton, Florida, 2005.

Kececioglu, D., *Maintainability, availability, and operational readiness engineering handbook*, vol. 1, DEStech Publications, Inc, Lancaster, Pennsylvania, United States, 2003.

Sharma, G. and Rai, R.N., Age based overhaul policy for multiple repairable systems with imperfect maintenance: Case study of aero engines. *Int. J. Math. Eng. Manag. Sci.*, 6, 1, 193, 2021a.

8

Study and Modeling of Factors Affecting the REI of Repairable Systems

8.1 Introduction

The concept of GRP involves a parameter called REI, denoted by "q," along with β and η, which represents the repair quality or repair effectiveness (RE). This quantitative parameter "q" acquired adequate consideration and is widely used in the literature. The estimation of REI is done with the help of field failure data. From Eq. (2.28), it can be noticed that as "q" moves toward 0, repair quality gets better or improves. Thus, the estimated "q" value from the field failure data plays a substantial role in understanding the quality of repair.

Simultaneously, the value of "q" reveals the notion of RE such that if q = 0.70, then RE would be 0.30 as shown in Eq. (8.1). Henceforth, RE represents the restoration percentage of the system after repair.

$$RE = 1 - REI(q) \tag{8.1}$$

Repair effectiveness could be dependent on numerous subjective factors and requires qualitative analysis for a better appreciation of repair effects and performance. The quantitative assessment of RE is necessary but not sufficient to analyze it completely. In the literature, adequate attention is given to the quantitative analysis of "q" and estimation of model parameters (η, β, and q) with the help of the maximum likelihood function, but work is limited toward the understanding of the physical inference of RE. This chapter reveals the drawbacks of the quantitative assessment of RE and emphasizes the necessity for further examining it qualitatively to substantiate its quantitative nature.

Garima Sharma and Rajiv Nandan Rai. Advanced Techniques for Maintenance Modeling and Reliability Analysis of Repairable Systems, (151–170) © 2023 Scrivener Publishing LLC

8.2 Limitations of the Quantitative Assessment of RE

With the help of the following example, we explain the limitations of the quantitative assessment of RE.

Example: Let us consider the model developed by Rai and Bolia (2015) which relates risk priority number (RPN) with "q" considering various FMs of an aero engine. The estimated performance parameters for the considered FMs in the study are presented in Table 8.1.

The RE value for FM1 (corrosion of compressor blades) is slightly better than that of FM3 (creep-fatigue fracture of turbine blades). The expected and estimated values of "q" along with RE for both FMs (R. N. Rai and Bolia, 2015) are tabulated in Table 8.2.

From Table 8.2, the authors state that if RE is increased from 0.15 to 0.80, there is a decrease of (0.8183 – 0.185 = 0.633) 66.5% in $F(t)$ for FM1. Similarly, for FM3, if RE is increased from 0.00 to 0.85, there is a decrease of (0.6817 – 0.1821 = 0.4996) 49.96% in $F(t)$. Afterward, the work (R. N. Rai and Bolia, 2015) ends up with some common recommendations to improve RE for all considered cases.

Although the above example shows a sensitivity between $F(t)$ and "q," it lacks in addressing the following aspects about $F(t)$ and the repair quality:

- What physical significance can be drawn from the suggested numerical values of RE?
- What could be the probable reasons for RE being good or bad?
- What measures should be adopted to improve RE and to what extent?
- What are the factors that can affect RE the most?

The drawbacks of the quantitative assessment of RE as discussed above can be effectively addressed by investigating the various factors affecting RE. The factors could be subjective with interdependencies among each other. The improvement in RE can be examined by understanding the implications of those factors and their connections with RE (Sharma and Rai, 2021). Moreover, the factors can be further categorized into several dependent subfactors. Thus, the modeling and analysis of interdependencies and the combined effect of those factors/subfactors on RE are required to be done.

Thus, the purpose of this chapter is to investigate various factors and model their interdependency to understand and establish their measured effects on RE, so that the required and fruitful actions can be initiated to

Table 8.1 Estimated parameters of the two failure modes of aero engine.

Aero engine failure modes	**Shape parameter (β)**	**Scale parameter (η)**	**REI (q)**	**RE = 1 − REI (q)**	**Failure probability $F(t)$**
Corrosion of compressor blades (FM1) (R. N. Rai and Bolia, 2015)	1.46	410.416 h	0.85	0.15	0.8183
Creep-fatigue fracture of turbine blades (FM3) (R. N. Rai and Bolia, 2015)	0.92	447.155 h	1.00	0.00	0.6817

Table 8.2 Expected RE values to decrease failure probability.

Aero engine failure modes	Estimated values			Expected values		
	REI (q)	**RE = 1 − REI (q)**	$F(t)$	**REI (q)**	**RE = 1 − REI (q)**	$F(t)$
FM1 (R. N. Rai and Bolia, 2015)	0.85	0.15	0.8183	0.20	0.80	0.185
FM3 (R. N. Rai and Bolia, 2015)	1.00	0.00	0.6817	0.15	0.85	0.1821

improve the quality of repair, especially by those industries which deal with the large, complex, and critical repairable systems. The measured effects of developed factors on RE and their interdependencies will lead to better insight to understand repair processes, and the needed remedial measures can be undertaken for overall improvement. Better repair quality will definitely enhance the life of the products and reduce the life cycle costs in the long run. The complete approach was to investigate factors/subfactors to develop a dependency model for RE to improve the overall repair process.

8.3 Investigation of Factor/Subfactors Affecting RE

This section explains the factors and subfactors that affect RE the most. The following 11 primary factors that affect RE are initially discussed along with their subfactors followed by a discussion on their interactions (Sharma and Rai, 2020a):

I. Human error (HE)
II. Repair cost (RC)
III. Preventive maintenance effectiveness (PME)
IV. Maintainability (MN)
V. Equipment effectiveness (EE)
VI. Skill (SK)
VII. Tools, testers, and ground equipment (TTGE)
VIII. Environmental condition (EC)
IX. Spare parts quality (SPQ)
X. Equipment complexity (EX)
XI. Task complexity (TX)

These 11 factors and their interdependencies are selected in a manner to provide a collective coverage of all possibilities that can affect the quality of repair. Each factor can be affected by its subfactors. The details of each factor and its subfactor(s) are presented in the subsequent paragraphs.

Human errors (HE): Human errors are most common in affecting the repair quality quite intensively (Rajiv Nandan Rai and Sharma, 2018). Many factors can cause HE, but in terms of human psychology (G. Peng, 2014), there might be two responsible subfactors that can generate human errors, such as (Rajiv Nandan Rai and Sharma, 2018; Sharma and Rai, 2021f):

a. Stress level (SL)
b. Supervision (SV)

If the stress level is very high, the quality of repair will be poor. Similarly, if supervision during the repair process is poor, the quality of repair is likely to degrade.

I. Repair cost (RC): If the cost of repair is very high, the organization may find it challenging to create proper repair facilities resulting in rendering poor quality of repair. Total repair cost is dependent on the following subfactors:
   a. Spare parts cost (SPC)
   b. Holding cost (HC)
   c. Setup cost (SUC)
   d. Admin cost (AC)
   e. Man-hour cost (MC)

   The above subfactors collectively can affect the total repair cost. Spare parts cost (procurement) and holding cost of spare parts are directly proportional to repair cost. Setup cost can get affected by TTGE required in the workplace. Administration, handling the repair activities, and the cost of the repair crew per hour also affect the total repair cost. If the above-described costs are optimal, then repair quality is expected to be good.
II. Preventive maintenance effectiveness (PME): Since preventive maintenance (PM) is performed to prevent failures, its quality plays an important role in deciding the RE. In other words, an effective PM results in a reduced number of failures leading to fewer interventions to the system.

Effective PM assists in reducing the wear out of the system and could be dependent on the following subfactors:

a. Documentation (DC)
b. Timely detection, analysis, and corrections (TDAC)
c. Standard preventive maintenance procedures (SPMP)

While performing PM, proper documentation of each activity, timely detection of snags, and corrections are necessary. In addition, the procedure followed for PM should be as per the standard operating procedures laid down by the original equipment manufacturer (OEM). Thus, SPMP can as well be affected by laid-down maintenance plans (MP).

III. Maintainability (MN): Maintainability is the probability to restore a failed system to an operating condition within a given time period by following standard procedures. Maintainability is considered to be a function of time and is dependent on the following subfactors:

a. Modularity (MD)
b. Analyzability (AS)
c. Repair scheme standardization (RSS)

If the system has a modular structure, the application of repair schemes is comparatively easier and less time-consuming. Modularity can be defined in terms of portability (PT) of the system, accessibility (ACS) to the defective part, and interchangeability (INT) of the components. Analyzability can be measured in terms of testability (TST) of the failed components. Standardization of repair schemes is also essential for good quality of repair. If the system maintainability is low, it would be difficult to perform repair tasks leading to poor quality of repair.

IV. Equipment effectiveness (EE): If the aging factor of a system is high, then the implemented repair actions will not be very effective. Hence, as the system gets older, the system's effectiveness decreases. The effectiveness of equipment also decreases because of the following subfactors:

a. Breakdown frequency (BDF)
b. Unscheduled maintenance (UM)

Moreover, the obsolescence (OBS) of the system can affect both BDF and UM. In addition, BDF causes frequent UMs.

If EE is high, then repair quality or RE could be much better.

V. Skill (SK): The skill of the crew performing the repair work is equally important. Underskilled workers performing poor quality of repair may lead to damages and further malfunctioning of the system. A skilled crew can be rated on the following subfactors:
   a. Experience (EXP)
   b. Appraisal (AP)
   c. Technical assistance (TA)
   d. Training (TN)

   The repair will be effective if the repair crew is selected as per the above-stated criteria. The education (EDU) of the repair crew, which is considered part of training, is equally essential.

VI. Tools, testers, and ground equipment (TTGE): The facilities provided by the organization for repair in terms of advanced repair equipment play a vital role to improve the RE. Test and ground equipment are covered under those facilities. The effectiveness of TTGE could be decided by considering various subfactors such as:
   a. Multifunctionality (MF)
   b. Portability (PT)
   c. Accuracy (AR)
   d. Standardization (STD)

   Multifunctionality is defined as the versatility of the testers to perform various functions. Portability is the easiness to carry to facilitate the testers to be used in inaccessible areas. The accuracy of TTGE is related to its calibration (CLB) and standardization signifying the readability (RED), annunciation (ANN), and availability of TTGE (TTGEA).

VII. Environmental condition (EC): Adverse environmental conditions may lead to crew discomfort and malfunctioning of equipment. Hence, the ECs created and provided by the organization to the repair crew affect the quality of repair to a great extent. The following subfactors are considered under EC:
   a. Ergonomics (ERG)
   b. Temperature and humidity (TAH)
   c. Personal protective equipment (PPE)

Correct ergonomics and comfortable TAH levels enable the repair crew to work peacefully and effectively. The effectiveness of personal protective equipment can be measured by its efficacy in protecting a person from the ill effects of noise, vibration (ENV), and chemical health hazards and radiations (CHHR). Thus, one of the causes of poor quality repair is unfavorable EC.

VIII. Spare parts quality (SPQ): The quality of repair can be affected to a great extent by the quality of spare parts used to replace the failed parts/components. Bad quality of spares can lead to low RE and frequent failures of the same component. The quality of spare parts can be checked through its standardization (STD), which ensures the supply chain (SC) facility, availability (AVA), and quality (QUA) of a particular spare part.

IX. Equipment complexity (EX): Quality of repair also depends on the complexity of the system. If a system consists of numerous subsystems or components within it, then the quality of repair is likely to decrease with the increased number of subsystems. Thus, the higher the value of complexity, the lower the RE.

X. Task complexity (TX): The complexity of the repair task which is to be performed on the system could be responsible for the bad quality of repair.

The above-explained factors can directly influence the repair quality. As explained earlier, these 11 factors are dependent on their respective subfactors and are also dependent on each other. This interdependency is required to be established properly to model the problem. Thus, a brief narrative is provided to understand the factors' interdependencies.

- Human errors (HEs) can be a common effect of low skill, inadequate test and ground equipment, high task complexity, low maintainability, and unfavorable environmental conditions.
- Preventive maintenance (PME) can get affected by inadequate TTGE, low-skilled workers, low maintainability of the system, and human errors.
- Equipment complexity can increase task complexity, and task complexity can affect the maintainability of the system.
- Low equipment effectiveness can be a result of low PM quality and maintainability.

- Adverse ECs can affect the efficiency of the test and ground equipment as well as spare parts effectiveness and adverse working conditions can lead to human errors.

8.4 Tool Chosen for the Analysis

The mathematical modeling of dependencies of these factors with each other and RE is quite complex because of their subjective nature. Thus, Bayesian network (BN) (Holmes, 2008) modeling could be a suitable candidate for the dependency modeling of such factors. Bayesian networks (BNs) are renowned for their capability to handle the dependencies among subjective variables/factors in various fields (LarrañAga *et al.*, 2013; Kar, Sharma, and Rai, 2022). Bayesian networks use directed acyclic graph (DAG) to represent the dependencies among the variables or factors (del Sagrado *et al.*, 2016). The graph contains arcs and nodes. Nodes and arcs represent variables and the relationship between variables, respectively. Missing arcs between two variables indicate no conditional dependency (Uusitalo, 2007) between variables. The variables may have one or more states. The belief in the states of a node gets updated as the belief of connected nodes changes. The basics of BN are adequately covered in the book by Holmes and Jain (Holmes, 2008).

The following example describes how the BN model can be used for analysis.

✓ Solved Example
Considering the factors discussed in Section 8.3:

(a) Define the states for each factor/subfactor
(b) Build a Bayesian network to show the dependencies
(c) Generate the prior probability data to feed into the network
(d) Analyze the results assuming that if RE is high (i.e., 1), what are the probabilities for different factors/subfactors of being favorable?

Solution:

(a) Two states for each factor/subfactor (favorable for RE and unfavorable for RE) are considered as described in Table 8.3.
(b) With the concept of BN, a DAG is created, with the help of experts, to analyze the effects of various factors/subfactors

Table 8.3 Defined states of each factor/subfactor.

SI no.	Factors/subfactors	State (favorable for RE)	State (not favorable for RE)
1.	HE	No	Yes
2.	SL	Low	High
3.	SV	High	Low
4.	RC	Optimal	Low
5.	SPC	Optimal	Low
6.	HC	Optimal	Low
7.	SUC	Optimal	Low
8.	AC	Optimal	Low
9.	MC	Optimal	Low
10.	TTGEC	Optimal	Low
11.	PME	High	Low
12.	DC	Good	Poor
13.	TDAC	Yes	No
14.	SPMP	Yes	No
15.	MP	Good	Poor
16.	MN	High	Low
17.	MD	High	Low
18.	AS	High	Low
19.	TST	High	Low
20.	RSS	Yes	No
21.	PT	High	Low
22.	ACS	High	Low
23.	INT	High	Low
24.	EE	High	Low

(*Continued*)

Table 8.3 Defined states of each factor/subfactor. (*Continued*)

SI no.	Factors/subfactors	State (favorable for RE)	State (not favorable for RE)
25.	BDF	Low	High
26.	UM	Low	High
27.	OBS	Low	High
28.	SK	High	Low
29.	EXP	High	Low
30.	AP	Good	Poor
31.	TA	Yes	No
32.	TN	Good	Poor
33.	EDU	High	Low
34.	TTGE	Adequate	Inadequate
35.	MF	Yes	No
36.	AR	High	Low
37.	STD	Yes	No
38.	CLB	Good	Poor
39.	RED	High	Low
40.	ANN	Good	Poor
41.	TTGEA	High	Low
42.	EC	Favorable	Unfavorable
43.	ERG	Comfort	Discomfort
44.	TAH	Low	High
45.	PPE	Yes	No
46.	CHHR	LOW	High
47.	SPQ	Specified	Not specified
48.	SC	Effective	Noneffective

(*Continued*)

Table 8.3 Defined states of each factor/subfactor. (*Continued*)

SI no.	Factors/subfactors	State (favorable for RE)	State (not favorable for RE)
49.	AVA	High	Low
50.	QUA	High	Low
51.	EX	Low	High
52.	TX	Low	High
53.	ENV	Low	High
54.	RE	High	Low

on RE as shown in Figure 8.1. The software called "SamIam" is used for preparing the Bayesian network.

(c) Prior probability data can then be generated with the help of industry experts. Furthermore, the data are converted into weights using pairwise comparison (Saaty, 2008). The normalized weights of factors/subfactors affecting a particular factor are given in Table 8.4.

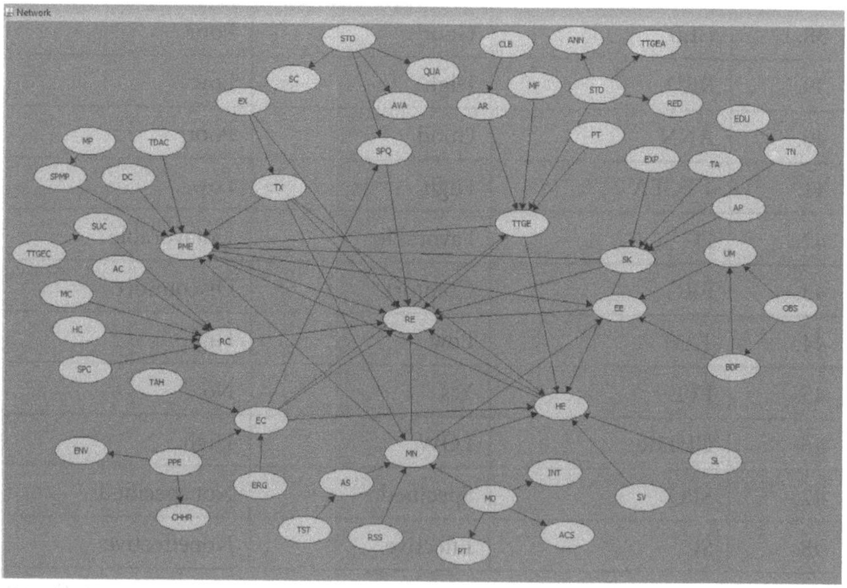

Figure 8.1 The BN model for factors affecting RE.

Table 8.4 Normalized weights obtained from pairwise comparison.

Factors	Affecting factors/subfactors	Normalized weights
HE	SK	0.349
	TTGE	0.109
	MN	0.237
	EC	0.050
	SL	0.037
	SV	0.139
	TX	0.079
RC	HC	0.111
	SPC	0.455
	SUC	0.111
	AC	0.057
	MC	0.264
PME	TTGE	0.140
	TX	0.106
	MN	0.314
	HE	0.076
	SK	0.239
	DC	0.029
	TDAC	0.054
	STD	0.041
MN	TX	0.169
	RSS	0.074
	MD	0.565
	AS	0.192

(Continued)

Table 8.4 Normalized weights obtained from pairwise comparison. (*Continued*)

Factors	Affecting factors/subfactors	Normalized weights
EE	PME	0.150
	MN	0.076
	BDF	0.514
	UM	0.260
SK	EXP	0.514
	AP	0.076
	TA	0.150
	TN	0.260
TTGE	EC	0.051
	MF	0.110
	PT	0.110
	AR	0.407
	STD	0.321
EC	TAH	0.638
	ERG	0.258
	PPE	0.105
SPQ	EC	0.400
	STD	0.600
EX	EX	1.000
TX	EX	1.000
RE	EE	0.108
	SK	0.248
	TTGE	0.077
	SPQ	0.152
	TX	0.029

(*Continued*)

Table 8.4 Normalized weights obtained from pairwise comparison. (*Continued*)

Factors	Affecting factors/subfactors	Normalized weights
	PME	0.056
	RC	0.036
	MN	0.167
	HE	0.044
	EC	0.019
	EX	0.062

(d) Allowing that RE is high (given evidence) in the model, the percentile effect of each factor on RE is observed. The results are shown in Figure 8.2 and tabulated in Table 8.5. The results provide insight into how RE can be significantly improved by treating the identified factors/subfactors in a proportion as acquired from the developed BN model.

A demographic view is also shown in Figure 8.3 considering the 11 primary factors. In the analysis, it is observed that EX is affecting RE the most since the complexity of the equipment plays a significant role during repair,

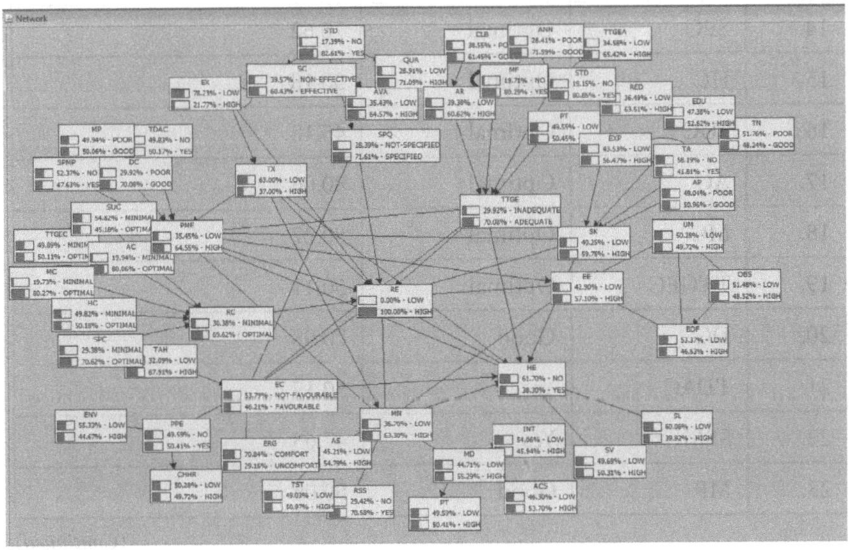

Figure 8.2 The Bayesian network model.

Table 8.5 Results using the BN model.

SI no.	Factors/ subfactors	Factorable state of factors/ subfactors	Probability of being factors/ subfactors favorable when RE is high (in %)
1.	HE	No	61.70
2.	RC	Optimal	69.62
3.	PME	High	64.55
4.	MN	High	63.30
5.	EE	High	57.10
6.	SK	High	59.75
7.	TTGE	Adequate	70.08
8.	EC	Favorable	46.21
9.	SPQ	Specified	71.61
10.	EX	Low	78.23
11.	TX	Low	63.00
12.	SL	Low	60.08
13.	SV	High	50.31
14.	SPC	Optimal	70.62
15.	HC	Optimal	50.18
16.	SUC	Optimal	45.18
17.	AC	Optimal	80.06
18.	MC	Optimal	80.27
19.	TTGEC	Optimal	50.11
20.	DC	Good	70.08
21.	TDAC	Yes	50.17
22.	SPMP	Yes	47.63
23.	MP	Good	50.06

(Continued)

Table 8.5 Results using the BN model. (*Continued*)

SI no.	Factors/ subfactors	Factorable state of factors/ subfactors	Probability of being factors/ subfactors favorable when RE is high (in %)
24.	MD	High	55.29
25.	AS	High	54.79
26.	RSS	Yes	70.58
27.	PT of equipment	High	50.41
28.	ACS	High	53.70
29.	INT	High	45.94
30.	TST	High	50.97
31.	BDF	Low	53.37
32.	UM	Low	50.28
33.	OBS	Low	51.48
34.	EXP	High	56.47
35.	AP	Good	50.96
36.	TA	Yes	41.81
37.	TN	Good	48.24
38.	EDU	High	52.62
39.	MF	Yes	80.29
40.	AR	High	60.62
41.	STD of TTGE	Yes	80.85
42.	CLB	Good	61.45
43.	PT of TTGE	High	50.45
44.	RED	High	63.51
45.	ANN	Good	71.59
46.	TTGEA	High	65.42

(*Continued*)

Table 8.5 Results using the BN model. (*Continued*)

SI no.	Factors/ subfactors	Factorable state of factors/ subfactors	Probability of being factors/ subfactors favorable when RE is high (in %)
47.	ERG	Comfort	70.84
48.	TAH	Low	32.09
49.	PPE	Yes	50.41
50.	ENV	Low	55.33
51.	CHHR	Low	50.28
52.	STD of Spare Parts	Yes	82.61
53.	SC	Effective	60.43
54.	AVA	High	64.57
55.	QUA	High	71.09

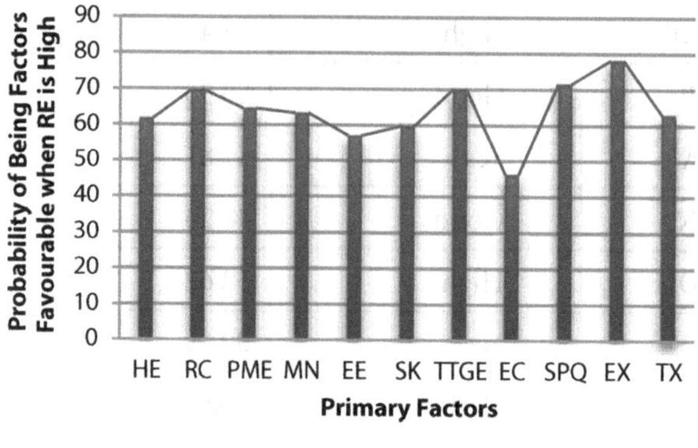

Figure 8.3 Analysis of the results obtained from the BN model.

followed by SPQ, TTGE, RC, PME, MN, TX, HE, SK, EE, and EC, respectively. If the system is easily maintainable or less critical, then it could be repaired effortlessly, but a complex system requires intense and specialized repair activities resulting in a decrease of RE.

Note: For a better understanding of the readers, the path from methodology to results is summarized and attached in Appendix E. In addition, the pairwise comparison matrices considering primary factors are also provided in Appendix E.

8.5 Chapter Summary

Understanding the physical significance of RE is unavoidable to analyze repair quality and seek remedial measures to improve the performability of a system. This chapter discusses the factors/subfactors that affect RE the most and emphasizes using BN to model the interdependencies between the factors.

The chapter creates a paradigm to extend future studies on other various factors affecting the RE of heavy industrial equipment. If the availability of such equipment is kept high, it will surely affect the productivity and throughput of the production lines which will witness a substantial improvement in the life cycle costs of the system and will enhance the understanding of the concept of quality of maintenance at MRO facilities.

❖Exercise Problem

Considering the BN network shown in the "Solved Example," analyze the probability of achieving high RE by giving the following evidence shown in Table 8.6. Provide evidence one after the other and observed the changes in RE after the given evidence (keep the previous evidence unchanged while giving a new one):

Table 8.6 List of evidence.

SI no.	Given evidence
1.	No HE
2.	PME is low
3.	ECs are favorable
4.	EXP of repair person is low
5.	SPC is optimal
6.	No STD of SPQ
7.	TTGE is STD
8.	BDF is high
9.	No RSS
10.	EX is low

References

Holmes, D.E., *Innovations in bayesian networks: Theory and applications*, vol. 156, Springer, Berlin, Germany, 2008.

Kar, A., Sharma, G., Rai, R.N., A Fuzzy bayesian network-based approach for modeling and analyzing factors causing process variability. *Int. J. Qual. Reliab. Manag.*, 40, 3, 820–44, 2022.

LarrañAga, P., Karshenas, H., Bielza, C., Santana, R., A review on evolutionary algorithms in bayesian network learning and inference tasks. *Inf. Sci.*, 233, 109–25, 2013.

Peng, G., *On human errors in maintenance: Risk potential and mitigation*, University of Stavanger, Stavanger, Norway, 2014.

Rai, R.N. and Bolia, N., Modified FMEA model with repair effectiveness factor using generalized renewal process. *SRESA J. Life Cycle Reliab. Saf. Eng.*, 4, 36–46, 2015.

Rai, R.N. and Sharma, G., Human reliability analysis technique selection for life support systems maintenance of orbital space stations using fuzzy AHP and ANN, in: *International Conference on Applied Human Factors and Ergonomics*, Springer, pp. 128–37, 2018.

Saaty, T.L., Decision making with the analytic hierarchy process. *Int. J. Serv. Sci.*, 1, 1, 83–98, 2008.

del Sagrado, J., Sanchez, J.A., Rodríguez, F., Berenguel, M., Bayesian networks for greenhouse temperature control. *J. Appl. Log.*, 17, 25–35, 2016.

Sharma, G. and Rai, R.N., Modeling and analysis of factors affecting repair effectiveness of repairable systems using bayesian network. *Appl. Soft Comput.*, 92, 106261, 2020a.

Sharma, G. and Rai, R.N., 10 - Methodology to select human reliability analysis technique for repairable systems, in: *Safety and Reliability Modeling and Its Applications*, H. Pham and M. Ram (Eds.), pp. 261–83, Elsevier, Amsterdam, Netherlands, 2021f, https://doi.org/10.1016/B978-0-12-823323-8.00007-6.

Sharma, G. and Rai, R.N., Modified repair quality index for complex repairable systems, in: *Proceedings of the 30th European Safety and Reliability Conference and the 15th Probabilistic Safety Assessment and Management Conference*, Research Publishing, Singapore, Venice, pp. 2851–58, 2021, https://doi.org/10.3850/978-981-14-8593-0.

Uusitalo, L., Advantages and challenges of bayesian networks in environmental modelling. *Ecol. Modell.*, 203, 3–4, 312–18, 2007.

Appendix A

NHPP MLEs

$$\frac{\partial Log(L)}{\partial \beta} = \frac{\sum_{l=1}^{K} p_l}{\beta} - log(\eta)\sum_{l=1}^{K} p_l + \sum_{l=1}^{K}\sum_{i=1}^{p_l}\sum_{i=1}^{p}\left[\left(\frac{t_{l,i-1}}{\eta}\right)^{\beta} log\left(\frac{t_{l,i-1}}{\eta}\right) - \left(\frac{t_{l,i}}{\eta}\right)^{\beta} log\left(\frac{t_{l,i}}{\eta}\right)\right] + \sum_{l=1}^{K}\sum_{i=1}^{p_l} log t_{li} = 0 \tag{A.1}$$

$$\frac{\partial Log(L)}{\partial \eta} = -\frac{\beta\sum_{l=1}^{K} p_l}{\eta} - \beta\eta^{-\beta-1}\sum_{l=1}^{K}\sum_{i=1}^{p_l}\left(t_{l,i-1}{}^{\beta} - t_{l,i}{}^{\beta}\right) = 0 \tag{A.2}$$

KI MLEs

$$\frac{\partial Log(L)}{\partial \beta} = \frac{\sum_{l=1}^{K} p_l}{\beta} - log(\eta)\sum_{l=1}^{K} p_l + \sum_{l=1}^{K}\left[\sum_{i=1}^{p_l}\left[\left(\frac{V_{l,i-1}}{\eta}\right)^{\beta} log\left(\frac{V_{l,i-1}}{\eta}\right) - \left(\frac{V_{l,i-1}+x_{li}}{\eta}\right)^{\beta} log\left(\frac{V_{l,i-1}+x_{li}}{\eta}\right)\right]\right] + \sum_{l=1}^{K}\left[\sum_{i=1}^{p_l} log(V_{l,i-1}+x_{l,i})\right] = 0 \tag{A.3}$$

$$\frac{\partial Log(L)}{\partial \eta} = -\frac{\beta\sum_{l=1}^{K} p_l}{\eta} - \beta\eta^{-\beta-1}\sum_{l=1}^{K}\sum_{i=1}^{p_l}\left[(V_{l,i-1})^{\beta} - (V_{l,i-1}+x_{li})^{\beta}\right] = 0 \tag{A.4}$$

$$\frac{\partial Log(L)}{\partial q} = (\beta - 1)\sum_{l=1}^{K}\sum_{i=1}^{p_l}\frac{\sum_{j=1}^{i-1} x_{lj}}{q\sum_{j=1}^{i-1} x_{lj} + x_{li}} + \frac{1}{\eta^{\beta}}\sum_{l=1}^{K}\sum_{i=1}^{p_l}\left[\beta\left(q\sum_{j=1}^{i-1} x_{lj}\right)^{\beta-1}\left(\sum_{j=1}^{i-1} x_{lj}\right) - \beta\left(q\sum_{j=1}^{i-1} x_{lj} + x_{li}\right)^{\beta-1}\left(\sum_{j=1}^{i-1} x_{lj}\right)\right] = 0 \quad \text{(A.5)}$$

KII MLEs

- β, η MLEs are same as Eq. (A.1) and (A.2).

$$\frac{\partial Log(L)}{\partial q} = (\beta - 1)\sum_{l=1}^{K}\sum_{i=1}^{p_l}\frac{\sum_{j=1}^{i}(i-j+1)q^{i-j}x_{lj}}{\sum_{j=1}^{i} q^{i-j+1}x_j} + \frac{1}{\eta^{\beta}}\sum_{l=1}^{K}\sum_{i=1}^{p_l}\left[\beta\left(\sum_{j=1}^{i-1} q^{i-j}x_{lj}\right)^{\beta-1}\left(\sum_{j=1}^{i-1}(i-j)q^{i-j-1}x_{lj}\right) - \beta\left(\sum_{j=1}^{i} q^{i-j+1}x_{lj}\right)^{\beta-1}\left(\sum_{j=1}^{i}(i-j+1)q^{i-j}x_{lj}\right)\right] = 0 \quad \text{(A.6)}$$

Appendix B

KI MLEs (Nasr *et al.* Model)

$$\frac{\partial Log(L)}{\partial \beta} = \sum_{l=1}^{K}\left[\sum_{i=1}^{n_l}\left[\left(\frac{V_{l,i-1}}{\eta}\right)^{\beta} log\left(\frac{V_{l,i-1}}{\eta}\right) - \left(\frac{(V_{li-1}+x_{li})}{\eta}\right)^{\beta} log\left(\frac{(V_{l,i-1}+x_{li})}{\eta}\right)\right]\right] + \frac{1}{\beta}\sum_{l=1}^{K}[\sum_{i=1}^{n_l}\delta_{li}] + \sum_{l=1}^{K}[\sum_{i=1}^{n_l}\delta_{li} log(V_{l,i-1}+x_{li})] - log(\eta)\sum_{l=1}^{K}[\sum_{i=1}^{n_l}\delta_{li}] = 0 \tag{B.1}$$

$$\frac{\partial Log(L)}{\partial \eta} = -\beta\eta^{-\beta-1} - \sum_{l=1}^{K}\left[\sum_{i=1}^{n_l}\left[\left(V_{l,i-1}\right)^{\beta} - \left(V_{l,i-1}+x_{li}\right)^{\beta}\right]\right] - \frac{1}{\beta}\sum_{l=1}^{K}[\sum_{i=1}^{n_l}\delta_{li}] = 0 \tag{B.2}$$

$$\frac{\partial Log(L)}{\partial q_{CM}} = (\beta-1)\sum_{l=1}^{K}\left[\sum_{i=1}^{n_l}\delta_{li}\frac{\sum_{j=1}^{i-1}\delta_{lj}q_{CM}^{\delta_{lj}-1}q_{SPM}^{1-\delta_{lj}}x_{l,j}}{\sum_{j=1}^{i-1}q_{CM}^{\delta_{lj}}q_{PM}^{1-\delta_{lj}}x_{lj}+x_{li}}\right] + \frac{1}{\eta^{\beta}}\sum_{l=1}^{K}\left[\sum_{i=1}^{n_l}\left[\beta\left(\sum_{j=1}^{i-1}\delta_{lj}q_{CM}^{\delta_{lj}-1}q_{SPM}^{1-\delta_{lj}}x_{lj}\right)\left(\sum_{j=1}^{i-1}q_{CM}^{\delta_{lj}}q_{SPM}^{1-\delta_{lj}}x_{lj}\right)^{\beta-1} - \beta\left(\sum_{j=1}^{i-1}\delta_{lj}q_{CM}^{\delta_{lj}-1}q_{SPM}^{1-\delta_{lj}}x_{lj}\right)\left(\sum_{j=1}^{i-1}q_{CM}^{\delta_{lj}}q_{SPM}^{1-\delta_{lj}}x_{lj}+x_{li}\right)^{\beta-1}\right]\right] = 0 \tag{B.3}$$

$$\frac{\partial Log(L)}{\partial q_{SPM}} = (\beta-1)\sum_{l=1}^{K}\left[\sum_{i=1}^{n_l}\delta_{li}\frac{\sum_{j=1}^{i-1}(1-\delta_{lj})q_{CM}^{\delta_{lj}}q_{SPM}^{-\delta_{lj}}x_{lj}}{\sum_{j=1}^{i-1}q_{CM}^{\delta_{lj}}q_{SPM}^{1-\delta_{lj}}x_{lj}+x_{li}}\right]$$
$$+\frac{1}{\eta^{\beta}}\sum_{l=1}^{K}\left[\sum_{i=1}^{n_l}\left[\beta\left(\sum_{j=1}^{i-1}(1-\delta_{lj})q_{CM}^{\delta_{lj}}q_{SPM}^{-\delta_{lj}}x_{lj}\right)\left(\sum_{j=1}^{i-1}q_{CM}^{\delta_{lj}}q_{SPM}^{1-\delta_{lj}}x_{lj}\right)^{\beta-1}\right.\right.$$
$$\left.\left.-\beta\left(\sum_{j=1}^{i-1}(1-\delta_{lj})q_{CM}^{\delta_{lj}}q_{SPM}^{-\delta_{lj}}x_{lj}\right)\left(\sum_{j=1}^{i-1}q_{CM}^{\delta_{lj}}q_{SPM}^{1-\delta_{lj}}x_{lj}+x_{li}\right)^{\beta-1}\right]\right]=0 \tag{B.4}$$

KII MLEs (Nasr *et al.* Model)

- β, η MLEs are same as Eq. (B.1) and (B.2)

$$\frac{\partial Log(L)}{\partial q_{CM}} = (\beta-1)\sum_{l=1}^{K}\left[\sum_{i=1}^{n_l}\delta_{li}\frac{\sum_{j=1}^{i}\{(i-j+1)\delta_{lj}\}q_{CM}^{\{(i-j+1)\delta_{lj}\}-1}q_{SPM}^{(i-j+1)(1-\delta_{lj})}x_{lj}}{\sum_{j=1}^{i}q_{CM}^{(i-j+1)\delta_{lj}}q_{SPM}^{(i-j+1)(1-\delta_{lj})}x_{lj}}\right]$$
$$+\frac{1}{\eta^{\beta}}\sum_{l=1}^{K}\left[\sum_{i=1}^{n_l}\left[\beta\left(\sum_{j=1}^{i-1}(i-j)\delta_{lj}q_{CM}^{\{(i-j)\delta_{lj}\}-1}q_{SPM}^{(i-j)(1-\delta_{lj})}x_{lj}\right)\right.\right.$$
$$\left(\sum_{j=1}^{i-1}q_{CM}^{(i-j)\delta_{lj}}q_{SPM}^{(i-j)(1-\delta_{lj})}x_{lj}\right)^{\beta-1}-\beta\left(\sum_{j=1}^{i}\left\{(i-j+1)\delta_{j}\right\}q_{CM}^{\{(i-j+1)\delta_{lj}\}-1}q_{SPM}^{(i-j+1)(1-\delta_{lj})}x_{lj}\right)$$
$$\left.\left.\left(\sum_{j=1}^{i}q_{CM}^{(i-j+1)\delta_{lj}}q_{SPM}^{(i-j+1)(1-\delta_{lj})}x_{lj}\right)^{\beta-1}\right]\right]=0 \tag{B.5}$$

$$\frac{\partial Log(L)}{\partial q_{SPM}} = (\beta-1)\sum_{l=1}^{K}\left[\sum_{i=1}^{n_l}\delta_{li}\frac{\sum_{j=1}^{i}\{(i-j+1)(1-\delta_{lj})\}q_{SPM}^{\{(i-j+1)(1-\delta_{lj})\}-1}q_{CM}^{(i-j+1)\delta_{lj}}x_{lj}}{\sum_{j=1}^{i}q_{CM}^{(i-j+1)\delta_{lj}}q_{SPM}^{(i-j+1)(1-\delta_{lj})}x_{lj}}\right]$$
$$+\frac{1}{\eta^{\beta}}\sum_{l=1}^{K}\left[\sum_{i=1}^{n_l}\left[\beta\left(\sum_{j=1}^{i-1}\left\{(i-j)(1-\delta_{j})\right\}q_{SPM}^{\{(i-j)(1-\delta_{lj})\}-1}q_{CM}^{(1-j)\delta_{lj}}x_{lj}\right)\right.\right.$$
$$\left(\sum_{j=1}^{i-1}q_{CM}^{(i-j)\delta_{lj}}q_{SPM}^{(i-j)(1-\delta_{lj})}x_{lj}\right)^{\beta-1}-\beta\left(\sum_{j=1}^{i}\left\{(i-j+1)(1-\delta_{j})\right\}q_{SPM}^{\{(i-j+1)(1-\delta_{lj})\}-1}q_{CM}^{(i-j+1)\delta_{lj}}x_{lj}\right)$$
$$\left.\left.\left(\sum_{j=1}^{i}q_{CM}^{(i-j+1)\delta_{lj}}q_{SPM}^{(i-j+1)(1-\delta_{lj})}x_{lj}\right)^{\beta-1}\right]\right]=0 \tag{B.6}$$

Appendix C

MATLAB Code for Imperfect Overhaul Model

```
clc;
close all;
clear all;
qc= xlsread('Random Parameters', sheet,'_:_'); % j
qp= xlsread('Random Parameters',sheet,'_:_'); % j
qo= xlsread('Random Parameters',sheet,'_:_'); % j
b= xlsread('Random Parameters',sheet,'_:_'); % j
for j=1:size(qc)
   Paranum=j;
 [~,sheet_name]=xlsfinfo('data.xlsx');
   for k=1:numel(sheet_name)
      data{k}=xlsread('System new data.xlsx',sheet_name{k});
   end
FirT=0; STerm=0; ThiT=0; FifT=0; ForT=0; SixT=0; Eta_Term1=0;
Eta_Term2=0;
 for l= 1:k
      sheet= l;
      t=data{1,l}(:,2); % Read TSN Value
      x=data{1,l}(:,3); % Read TBF (i)
      p=data{1,l}(:,4); % Read Delta value for each data point
      h=data{1,l}(:,5); % Overhaul term
      % ---------------------Main Calculation--------------------------
     T= 1500; del=0;sum1=0; sum=0;sum2=0; Overhltrm=0;n = numel(x);
     xo=750;Overhsum=0;
      %T=3000;del=0;sum1=0;sum=0;sum2=0;n=numel(x);vo=0;xm=1000;
         for m=1:size(p)
         del= del+p(m);
      end
```

```
% --------------------KI calculation------------------------------------------
 for i=2:size(x)
        vm(1)=0;
         vm(i)=vm(i-1);
          if h(i)==1
              vm(i)=vm(i-1)+(qo(j)*xm);
          v(i)=vm(i);
        else
 v(1)= x(1)*((qc(1))^(p(1))*(((qp(1))^(1-h(1)))^(1-p(1))));
 v(i)=((v(i-1)+(qc(j)*x(i)))^(p(i)))*((((v(i-1)+(qp(j)*x(i)))^(1-h(i))))^(1-p(i)));
  end
  Virtual_age = transpose(v);
  SecT1 =((((v(i-1))^(b(j)))-(((x(i))+(v(i-1)))^(b(j))))*(1-h(i)));
  sum1= sum1+SecT1;
  Ovrh=((((vm(i-1))^(b(j)))-(((xm)+(vm(i-1)))^(b(j))))*(h(i)));
  Overhltrm=Overhltrm+Ovrh;
  sum= sum+(p(i)*log((x(i))+(v(i-1))));
  SecT1_for_eta =(((((x(i))+(v(i-1)))^(b(j)))-((v(i-1))^(b(j))))*(1-h(i)));
  sum2=sum2+SecT1_for_eta ;
  Ovrh_for_eta=(((((xo)+(vo(i-1)))^(b(j)))-((vo(i-1))^(b(j))))*(h(i)));
  Overhsum= Overhsum + Ovrh_for_eta;
 end
  TimeT=((((v(n))^(b(j)))-((T-t(n)+(v(n)))^(b(j)))));
  TimeT_for_eta=(((T-t(n)+(v(n)))^(b(j)))-((v(n))^(b(j))));
    FirT = FirT+TimeT;
    STerm = STerm + sum1;
    ThiT= ThiT+del;
   ForT= ForT+del;
   FifT = FifT+sum;
   SixT=SixT+Overhltrm;
    Eta_Term1= Eta_Term1+sum2;
    Eta_Term2= Eta_Term2+Overhsum;
   Weibull_term= (a(j)*b(j)*(t(1))^(b(j)-1))*exp(-a(j)((t(1))^b(j)));
 end
   eta(j)=(((Eta_Term1+ Eta_Term2+TimeT_for_eta)/ForT)^(1/b(j)));
   a(j)=(1/(eta(j)^b(j)));
   a_value (j,:)= a(j);
   lnL_Time_Truncated(j,:)=Weibull_term+(a(j)*FirT)+(log(b(j))*ForT)+
   (((b(j))-1)*FifT) +(log(a(j))*ThiT)+ (a(j)*STerm)+(a(j)*SixT);
end
```

Appendix D

MLEs for the KI Censoring Model:

$$\frac{\partial Log(L)}{\partial \beta} = \frac{1}{\beta}\sum_{l=1}^{K}\left[\sum_{i=1}^{n_l}(\delta_{li} - \delta_{li}\varphi_{li})\right]$$
$$+\sum_{l=1}^{K}\left[\sum_{i=1}^{n_l}(\delta_{li} - \delta_{li}\varphi_{li})log\left(V_{l,i-1} + x_{li}\right)\right] - log(\eta)\sum_{l=1}^{K}\left[\sum_{i=1}^{n_l}(\delta_{li} - \delta_{li}\varphi_{li})\right]$$
$$+\sum_{l=1}^{K}\left[\sum_{i=1}^{n_l}(1-\varphi_{li})\left[\left(\frac{V_{l,i-1}}{\eta}\right)^{\beta} log\left(\frac{V_{l,i-1}}{\eta}\right) - \left(\frac{(V_{l,i-1} + x_{li})}{\eta}\right)^{\beta} log\left(\frac{(V_{l,i-1} + x_{li})}{\eta}\right)\right]\right]$$
$$+\sum_{l=1}^{K}\left[\sum_{i=1}^{n_l}(\varphi_{li})\left[\left(\frac{V_{l,i-1}}{\eta}\right)^{\beta} log\left(\frac{V_{l,i-1}}{\eta}\right) - \left(\frac{(V_{l,i-1} + x_{li})}{\eta}\right)^{\beta} log\left(\frac{(V_{l,i-1} + x_{li})}{\eta}\right)\right]\right] = 0 \tag{D.1}$$

$$\frac{\partial Log(L)}{\partial \eta} = -\frac{\beta}{\eta}\sum_{l=1}^{K}\left[\sum_{i=1}^{n_l}(\delta_{li} - \delta_{li}\varphi_{li})\right] - \beta\eta^{-\beta-1}$$
$$\sum_{l=1}^{K}\left[\sum_{i=1}^{n_l}(1-\varphi_{li})\left[\left(V_{l,i-1}\right)^{\beta} - \left(V_{l,i-1} + x_{li}\right)^{\beta}\right]\right] - \beta\eta^{-\beta-1} \tag{D.2}$$
$$\sum_{l=1}^{K}\left[\sum_{i=1}^{n_l}(\varphi_{li})\left[\left(V_{l,i-1}\right)^{\beta} - \left(V_{l,i-1} + x_{li}\right)^{\beta}\right]\right] = 0$$

$$\frac{\partial Log(L)}{\partial q_{CM}} = (\beta-1)\sum_{l=1}^{K}\left[\sum_{i=1}^{n_l}(\delta_{li}-\delta_{li}\varphi_{li})\frac{\sum_{j=1}^{i-1}\{(1-\varphi_{lj})\delta_{lj}\}q_{CM}^{\{(1-\varphi_{lj})\delta_{lj}\}-1}q_{SPM}^{(1-\varphi_{lj})(1-\delta_{lj})}x_{lj}}{\sum_{j=1}^{i-1}q_{CM}^{(1-\varphi_{lj})\delta_j}q_{SPM}^{(1-\varphi_{lj})1-\delta_{lj}}x_{lj}+x_{li}}\right]$$
$$+\frac{1}{\eta^{\beta}}\sum_{l=1}^{K}\left[\sum_{i=1}^{n_l}(1-\varphi_{li})\left[\beta\left(\sum_{j=1}^{i-1}\{(1-\varphi_{lj})\delta_{lj}\}q_{CM}^{\{(1-\varphi_{lj})\delta_{lj}\}-1}q_{SPM}^{(1-\varphi_{lj})(1-\delta_{lj})}x_{lj}\right)\right.\right.$$
$$\left(\sum_{j=1}^{i-1}q_{CM}^{(1-\varphi_{lj})\delta_{lj}}q_{SPM}^{(1-\varphi_{lj})1-\delta_{lj}}x_{lj}\right)^{\beta-1}-\beta\left(\sum_{j=1}^{i-1}\left\{\left(1-\varphi_{lj}\right)\delta_{lj}\right\}q_{CM}^{\{(1-\varphi_{lj})\delta_{lj}\}-1}q_{SPM}^{(1-\varphi_{lj})(1-\delta_{lj})}x_{lj}\right)$$
$$\left.\left.\left(\sum_{j=1}^{i-1}q_{CM}^{(1-\varphi_{lj})\delta_{lj}}q_{SPM}^{(1-\varphi_{lj})1-\delta_{lj}}x_{lj}+x_{li}\right)^{\beta-1}\right]\right]+\frac{1}{\eta^{\beta}}\sum_{l=1}^{K}\left[\sum_{i=1}^{n_l}(\varphi_{li})\left[\beta\left(\sum_{j=1}^{i-1}\left\{\left(1-\varphi_{lj}\right)\delta_{lj}\right\}\right.\right.\right.$$
$$\left.q_{CM}^{\{(1-\varphi_{lj})\delta_{lj}\}-1}q_{SPM}^{(1-\varphi_{lj})(1-\delta_{lj})}x_{lj}\right)\left(\sum_{j=1}^{i-1}q_{CM}^{(1-\varphi_{lj})\delta_{lj}}q_{SPM}^{(1-\varphi_{lj})1-\delta_{lj}}x_{lj}\right)^{\beta-1}$$
$$-\beta(1\left(\sum_{j=1}^{i-1}\{(1-\varphi_{lj})\delta_{lj}\}q_{CM}^{\{(1-\varphi_{lj})\delta_{lj}\}-1}q_{SPM}^{(1-\varphi_{lj})(1-\delta_{lj})}x_{lj}\right)$$
$$\left.\left.\left(\sum_{j=1}^{i-1}q_{CM}^{(1-\varphi_{lj})\delta_{lj}}q_{SPM}^{(1-\varphi_{lj})1-\delta_{lj}}x_{lj}+x_{li}\right)^{\beta-1}\right]\right]=0 \tag{D.3}$$

$$\frac{\partial Log(L)}{\partial q_{SPM}} = (\beta-1)$$
$$\sum_{l=1}^{K}\left[\sum_{i=1}^{n_l}(\delta_{li}-\delta_{li}\varphi_{li})\frac{\sum_{j=1}^{i-1}\{(1-\varphi_{lj})(1-\delta_{lj})\}q_{SPM}^{\{(1-\varphi_{lj})(1-\delta_{lj})\}-1}q_{CM}^{(1-\varphi_{lj})(1-\delta_{lj})}x_{lj}}{\sum_{j=1}^{i-1}q_{CM}^{(1-\varphi_{lj})\delta_{lj}}q_{SPM}^{(1-\varphi_{lj})1-\delta_{lj}}x_{lj}+x_{li}}\right]$$
$$+\frac{1}{\eta^{\beta}}\sum_{l=1}^{K}\left[\sum_{i=1}^{n_l}(1-\varphi_{li})\left[\beta\left(\sum_{j=1}^{i-1}\{(1-\varphi_{lj})(1-\delta_{lj})\}\right.\right.\right.$$
$$\left.q_{SPM}^{\{(1-\varphi_{lj})(1-\delta_{lj})\}-1}q_{CM}^{(1-\varphi_{lj})\delta_{lj}}x_{lj}\right)\left(\sum_{j=1}^{i-1}q_{CM}^{(1-\varphi_{lj})\delta_{lj}}q_{SPM}^{(1-\varphi_{lj})1-\delta_{lj}}x_{lj}\right)^{\beta-1}$$
$$-\beta\left(\sum_{j=1}^{i-1}\left\{\left(1-\varphi_{lj}\right)(1-\delta_{lj})\right\}q_{SPM}^{\{(1-\varphi_{lj})(1-\delta_{lj})\}-1}q_{CM}^{(1-\varphi_{lj})\delta_{lj}}x_{lj}\right)$$
$$\left.\left.\left(\sum_{j=1}^{i-1}q_{CM}^{(1-\varphi_{lj})\delta_{lj}}q_{SPM}^{(1-\varphi_{lj})1-\delta_{lj}}x_{lj}+x_{li}\right)^{\beta-1}\right]\right]+\frac{1}{\eta^{\beta}}$$
$$\sum_{l=1}^{K}\left[\sum_{i=1}^{n_l}(\varphi_{li})\left[\beta\left(\sum_{j=1}^{i-1}\left\{\left(1-\varphi_{lj}\right)(1-\delta_{lj})\right\}\right.\right.\right.$$
$$\left.q_{SPM}^{\{(1-\varphi_{lj})(1-\delta_{lj})\}-1}q_{CM}^{(1-\varphi_{lj})\delta_{lj}}x_{lj}\right)\left(\sum_{j=1}^{i-1}q_{CM}^{(1-\varphi_{lj})\delta_{lj}}q_{SPM}^{(1-\varphi_{lj})1-\delta_{lj}}x_{lj}\right)^{\beta-1}$$
$$-\beta\left(\sum_{j=1}^{i-1}\left\{\left(1-\varphi_{lj}\right)(1-\delta_{lj})\right\}q_{SPM}^{\{(1-\varphi_{lj})(1-\delta_{lj})\}-1}q_{CM}^{(1-\varphi_{lj})\delta_{lj}}x_{lj}\right)$$
$$\left.\left.\left(\sum_{j=1}^{i-1}q_{CM}^{(1-\varphi_{lj})\delta_{lj}}q_{SPM}^{(1-\varphi_{lj})1-\delta_{lj}}x_{lj}+x_{li}\right)^{\beta-1}\right]\right]=0 \tag{D.4}$$

MLEs for the KII Censoring Model:

➤ β, η MLEs are same as Eq. (D.1) and (D.2)

$$\frac{\partial Log(L)}{\partial q_{CM}} = (\beta-1)\sum_{l=1}^{K}\left[\sum_{i=1}^{n_l}(\delta_{li}-\delta_{li}\varphi_{li}) \frac{\sum_{j=1}^{i-1}\{(1-\varphi_{lj})(i-j-\zeta_{l,i-1})\delta_{lj}\}q_{CM}^{\{(1-\varphi_{lj})(i-j-\zeta_{l,i-1})\delta_{lj}\}-1}q_{SPM}^{(1-\varphi_{lj})(i-j-\zeta_{l,i-1})(1-\delta_{lj})}x_{lj}}{\sum_{j=1}^{i-1}q_{CM}^{(1-\varphi_{lj})(i-j-\zeta_{l,i-1})\delta_{lj}}q_{SPM}^{(1-\varphi_{lj})(i-j-\zeta_{l,i-1})(1-\delta_{lj})}x_{lj}+x_{li}}\right]$$

$$+\frac{1}{\eta^{\beta}}\sum_{l=1}^{K}\left[\sum_{i=1}^{n_l}(1-\varphi_{li})\left[\beta\left(\sum_{i=1}^{n_l}\{(1-\varphi_{lj})(i-j-\zeta_{l,i-1})\delta_{lj}\}\right.\right.\right.$$
$$\left.q_{CM}^{\{(1-\varphi_{lj})(i-j-\zeta_{l,i-1})\delta_{lj}\}-1}q_{SPM}^{(1-\varphi_{lj})(i-j-\zeta_{l,i-1})(1-\delta_{lj})}x_{lj}\right)$$
$$\left(\sum_{j=1}^{i-1}q_{CM}^{(1-\varphi_{lj})(i-j-\zeta_{l,i-1})\delta_{lj}}q_{SPM}^{(1-\varphi_{lj})(i-j-\zeta_{l,i-1})(1-\delta_{lj})}x_{lj}\right)^{\beta-1}$$
$$-\beta\left(\sum_{j=1}^{i-1}\{(1-\varphi_{lj})(i-j-\zeta_{l,i-1})\delta_{lj}\}q_{CM}^{\{(1-\varphi_{lj})(i-j-\zeta_{l,i-1})\delta_{lj}\}-1}q_{SPM}^{(1-\varphi_{lj})(i-j-\zeta_{l,i-1})(1-\delta_{lj})}x_{lj}\right)$$
$$\left.\left.\left(\sum_{j=1}^{i-1}q_{CM}^{(1-\varphi_{lj})(i-j-\zeta_{l,i-1})\delta_{lj}}q_{SPM}^{(1-\varphi_{lj})(i-j-\zeta_{l,i-1})(1-\delta_{lj})}x_{lj}+x_{li}\right)^{\beta-1}\right]\right]+\frac{1}{\eta^{\beta}}$$

$$\sum_{l=1}^{K}\left[\sum_{i=1}^{n_l}(\varphi_{li})\left[\beta\left(\sum_{j=1}^{i-1}\{(1-\varphi_{lj})(i-j-\zeta_{l,i-1})\delta_{lj}\}\right.\right.\right.$$
$$\left.q_{CM}^{\{(1-\varphi_{lj})(i-j-\zeta_{l,i-1})\delta_{lj}\}-1}q_{SPM}^{(1-\varphi_{lj})(i-j-\zeta_{l,i-1})(1-\delta_{lj})}x_{lj}\right)$$
$$\left(\sum_{j=1}^{i-1}q_{CM}^{(1-\varphi_{lj})(i-j-\zeta_{l,i-1})\delta_{lj}}q_{SPM}^{(1-\varphi_{lj})(i-j-\zeta_{l,i-1})(1-\delta_{lj})}x_{lj}\right)^{\beta-1}$$
$$-\beta\left(\sum_{j=1}^{i-1}\{(1-\varphi_{lj})(i-j-\zeta_{l,i-1})\delta_{lj}\}q_{CM}^{\{(1-\varphi_{lj})(i-j-\zeta_{l,i-1}))\delta_{lj}\}-1}q_{SPM}^{(1-\varphi_{lj})(i-j-\zeta_{l,i-1})(1-\delta_{lj})}x_{lj}\right)$$
$$\left.\left.\left(\sum_{j=1}^{i-1}q_{CM}^{(1-\varphi_{lj})(i-j-\zeta_{l,i-1})\delta_{lj}}q_{SPM}^{(1-\varphi_{lj})(i-j-\zeta_{l,i-1})(1-\delta_{lj})}x_{lj}+x_{li}\right)^{\beta-1}\right]\right]=0 \tag{D.5}$$

$$
\begin{aligned}
&\frac{\partial Log(L)}{\partial q_{SPM}} = (\beta-1)\sum_{l=1}^{K}\Bigg[\sum_{i=1}^{n_l}(\delta_{li}-\delta_{li}\varphi_{li})\\
&\Bigg[\frac{\sum_{j=1}^{i-1}\{(1-\varphi_{lj})(i-j-\zeta_{l,i-1})(1-\delta_{lj})\}q_{SPM}^{\{(1-\varphi_{lj})(i-j-\zeta_{l,i-1})(1-\delta_{lj})\}-1}q_{CM}^{(1-\varphi_{lj})(i-j-\zeta_{l,i-1})\delta_{lj}}x_{lj}}{\sum_{j=1}^{i-1}q_{CM}^{(1-\varphi_{lj})(i-j-\zeta_{l,i-1})\delta_{lj}}q_{SPM}^{(1-\varphi_{lj})(i-j-\zeta_{l,i-1})(1-\delta_{lj})}x_{lj}+x_{li}}\Bigg]\\
&+\frac{1}{\eta^{\beta}}\sum_{l=1}^{K}\Bigg[\sum_{i=1}^{n_l}(1-\varphi_{li})\Bigg[\beta\Big(\sum_{j=1}^{i-1}\{(1-\varphi_{lj})(i-j-\zeta_{l,i-1})(1-\delta_{lj})\}\\
&q_{SPM}^{\{(1-\varphi_{lj})(i-j-\zeta_{l,i-1})(1-\delta_{lj})\}-1}q_{CM}^{(1-\varphi_{lj})(i-j-\zeta_{l,i-1})\delta_{lj}}x_{lj}\Big)\\
&\Big(\sum_{j=1}^{i-1}q_{CM}^{(1-\varphi_{lj})(i-j-\zeta_{l,i-1})\delta_{lj}}q_{SPM}^{(1-\varphi_{lj})(i-j-\zeta_{l,i-1})(1-\delta_{lj})}x_{lj}\Big)^{\beta-1}\\
&-\beta\Big(\sum_{j=1}^{i-1}\{(1-\varphi_{lj})(i-j-\zeta_{l,i-1})(1-\delta_{lj})\}q_{SPM}^{\{(1-\varphi_{lj})(i-j-\zeta_{l,i-1})(1-\delta_{lj})\}-1}q_{CM}^{(1-\varphi_{lj})(i-j-\zeta_{l,i-1})\delta_{lj}}x_{lj}\Big)\\
&\Big(\sum_{j=1}^{i-1}q_{CM}^{(1-\varphi_{lj})(i-j-\zeta_{l,i-1})\delta_{lj}}q_{SPM}^{(1-\varphi_{lj})(i-j-\zeta_{l,i-1})(1-\delta_{lj})}x_{lj}+x_{li}\Big)^{\beta-1}\Bigg]\Bigg]+\frac{1}{\eta^{\beta}}\\
&\sum_{l=1}^{K}\Bigg[\sum_{i=1}^{n_l}(\varphi_{li})\Bigg[\beta\Big(\sum_{j=1}^{i-1}\{(1-\varphi_{lj})(i-j-\zeta_{l,i-1})(1-\delta_{lj})\}\Big)\\
&q_{SPM}^{\{(1-\varphi_{lj})(i-j-\zeta_{l,i-1})(1-\delta_{lj})\}-1}q_{CM}^{(1-\varphi_{lj})(i-j-\zeta_{l,i-1})\delta_{lj}}x_{lj}\Big)\\
&\Big(\sum_{j=1}^{i-1}q_{CM}^{(1-\varphi_{lj})(i-j-\zeta_{l,i-1})\delta_{lj}}q_{SPM}^{(1-\varphi_{lj})(i-j-\zeta_{l,i-1})(1-\delta_{lj})}x_{lj}\Big)^{\beta-1}\\
&-\beta\Big(\sum_{j=1}^{i-1}\{(1-\varphi_{lj})(i-j-\zeta_{l,i-1})(1-\delta_{lj})\}q_{SPM}^{\{(1-\varphi_{lj})(i-j-\zeta_{l,i-1})(1-\delta_{lj})\}-1}q_{CM}^{(1-\varphi_{lj})(i-j-\zeta_{l,i-1})(1-\delta_{lj})}x_{lj}\Big)\\
&\Big(\sum_{j=1}^{i-1}q_{CM}^{(1-\varphi_{lj})(i-j-\zeta_{l,i-1})\delta_{lj}}q_{SPM}^{(1-\varphi_{lj})(i-j-\zeta_{l,i-1})(1-\delta_{lj})}x_{lj}+x_{li}\Big)^{\beta-1}\Bigg]\Bigg]=0
\end{aligned}
\tag{D.6}
$$

Code for Censoring Model

```
clc;
close all;
clear all;
sheet= 1;
% ------------------------New model--------------------------
qc= xlsread('Random Parameters', sheet,'_:_'); % j
qp= xlsread('Random Parameters',sheet,'_:_'); % j
% qo= xlsread('Random Parameters',sheet,'_:_'); % j
b= xlsread('Random Parameters',sheet,'_:_');  % j
```

```
for j=1:size(qc)
   Paranum=j
 [~,sheet_name]=xlsfinfo('Censoring_Data_FM1_(comp_blade).xlsx');
for k=1:numel(sheet_name)
 data{k}=xlsread('Censoring_Data_FM1_(comp_blade).xlsx',sheet_name{k});
      try
         % The statements you wish to try
      catch exception
      continue
         % What to do if an error occurs
      end
end
FirT =0; STerm1=0; ThiT =0; FifT =0; ForT =0; SixT=0; STerm2=0; Eta_
Term1=0; Eta_Term2=0;
 for l= 1:k-1
sheet= l;
t=data{1,l}(:,1);% Read TSN Value
x=data{1,l}(:,2);% Read TBF (i)
p=data{1,l}(:,3); % Read Delta value for each data point
L=data{1,l}(:,4);% Censoring
z=data{1,l}(:,5);% Censoring
% --------------------Main Calculation-------------------------
T= 550; del=0;sum1=0; sum=0;sum2=0; n = numel(x);sum3=0;sum4=0;
      for m=1:size(p)
        del= del+(p(m)-(p(m)*L(m)));
      end
% --------------------KI calculation---------------------------------------------
for i=2:size(x)
       v(1)= x(1)*(((qc(j))^(p(1))*(qp(j))^(1-p(1)))^(1-L(1)));
       v(i)= v(i-1)+ (x(i)*((qc(j))^(p(i))*(qp(j))^(1-p(i)))^(1-L(i)));
       Virtual_age = transpose(v);
       SecT1 =((v(i-1))^(b(j))-((x(i))+(v(i-1)))^(b(j)));
       sum1= sum1+((1-L(i))*SecT1);
       sum2=sum2+(L(i)*SecT1);
       sum= sum+((p(i)-(p(i)*L(i)))*log((x(i))+(v(i-1))));
  SecT1_for_eta1 =(((((x(i))+(v(i-1)))^(b(j)))-((v(i-1))^(b(j))))*(1-L(i)));
  sum3=sum3+SecT1_for_eta1 ;
  SecT1_for_eta2=(((((x(i))+(v(i-1)))^(b(j)))-((v(i-1))^(b(j))))*(L(i)));
  sum4=sum4+SecT1_for_eta2 ;
 end
 %
```

```
% -------------- KII calculation ------------------------------
% for i=2:size(x)
%              v(1)= x(1)*((((qc(j))^(p(1))*(qp(j))^(1-p(1))))^(1-L(1)));
%              k2=0;
%                  for r=1:i
% u(r)=((qc(j)^((i-r+1-z(i))*p(r))*qp(j)^((i-r+1-z(i))*(1-p(r))))^(1-
  L(r)))*x(r);
%                     k2=k2+u(r);
%                  end
%                   v(i,:)=transpose(k2);
%                SecT1 =((v(i-1))^(b(j))-((x(i))+(v(i-1)))^(b(j)));
%                sum1= sum1+((1-L(i))*SecT1);
%                sum2=sum2+(L(i)*SecT1);
%                sum= sum+((p(i)-(p(i)*L(i)))*log((x(i))+(v(i-1))));
% SecT1_for_eta1 =(((((x(i))+(v(i-1)))^(b(j)))-((v(i-1))^(b(j))))*(1-L(i)));
% sum3=sum3+SecT1_for_eta1 ;
% SecT1_for_eta2 =(((((x(i))+(v(i-1)))^(b(j)))-((v(i-1))^(b(j))))*(L(i)));
% sum4=sum4+SecT1_for_eta2 ;
% end
% -----------------------------------------------------------------

 TimeT=((((v(n))^(b(j)))-((T-t(n)+(v(n)))^(b(j)))));
 TimeT_for_eta=(((T-t(n)+(v(n)))^(b(j)))-((v(n))^(b(j))));
  FirT = FirT+TimeT;
  STerm1 = STerm1 + sum1;
  STerm2 = STerm2 + sum2;
  ThiT= ThiT+del;
  ForT= ForT+del;
  FifT = FifT+sum;
 Eta_Term1= Eta_Term1+sum3;
 Eta_Term2= Eta_Term2+sum4;
Weibull_term= (a(j)*b(j)*(t(1))^(b(j)-1))*exp(-a(j)((t(1))^b(j)));
 end
 eta(j)=(((Eta_Term1+ Eta_Term2+TimeT_for_eta)/ForT)^(1/b(j)));
 a(j)=(1/(eta(j)^b(j)));
  a_value (j,:)= a(j);
 lnL_Time_Truncated(j,:)=Weibull_term+(a(j)*FirT)+(a(j)*STerm1)+
 (a(j)*STerm2)+(log(a(j))*ThiT)+(log(b(j))*ForT)+(((b(j))-1)*FifT);
end
```

Appendix E

Steps of the Methodology to reproduce the Results for Better Comprehension of the Readers

The results obtained from the methodology presented in Chapter 8 of the book can be reproduced by adhering to the steps provided below:

Step 1: Investigation of factors/sub-factors
The factors, sub-factors, and their dependency structure considered in the methodology are investigated with the help of literature survey and industry experts. However, one can modify the network by choosing some important factors/sub-factors from the factors according to the organizational requirement. For example, if one does not want to include the factor "Maintainability (MN)" then it can be removed and analysis of repair effectiveness can be conducted by considering remaining factors.

Step 2: Formation of Bayesian network

- Considering the factors and sub-factors, a Bayesian network as shown in Figure 8.1 of the thesis, is formed using SamIam software in the paper. However, other off-the-shelf softwares like UnBBayesetc. can also be used to form such networks.
- The next step is to assign states to each factor/sub-factor in the software. We have assigned two states for each factor/sub-factor as explained at Table 8.3 of the thesis. One can assign more than two states depending on the requirements of the analysis. For example, for factor MN, we have selected two states i.e. low and high. However, one can assign two more states i.e. very low, and very high as well.

Step 3: Estimation of normalized input weights
We have used Analytical Hierarchy process (AHP) (Saaty 2008) to estimate the normalized weights for each factor/sub-factor but, other methods can also be used for the same. The method for estimating final weights is demonstrated considering one factor i.e., "SKILL" in the following example for the ready reference of the readers.

Example: The sub-factors assigned to SKILL are: Experience (EXP), Appraisal (AP), Technical Assistance (TA), and Training (TN).

- Prepare a pairwise comparison matrix based on Saaty scale
  With help of experts, we conduct the pairwise comparison as placed subsequently.
- Calculate the nth root of each sub-factor:

nth root of EX = $(1+5+3+3)^{\frac{1}{4}} = 2.59002$
n^{th} root of AP = 0.384162
n^{th} root of TA = 0.756027
n^{th} root of TN = 1.312771

- Calculate the final (normalized) weight of each sub-factor:

$$\text{Final weight of EX} = \frac{2.59002}{(2.59002 + 0.384162 + 0.756027 + 1.312771)}$$

$$= 0.513589$$

Final weight of AP = 0.076178
Final weight of TA = 0.149917
Final weight of TN = 0.260317

- Calculate and check the consistency ratio (CR)
  - Calculate λ and for each sub-factor:

$$\lambda\,(EX) = \frac{(1 \times 0.513589 + 5 \times 0.076178 + 3 \times 0.149917 + 3 \times 0.260317)}{0.513589}$$

$$= 4.137892844$$

λ (AP) = 4.125523341
λ (TA) = 4.227936947
λ (TN) = 4.256677912
$\lambda_{mean} = 4.187007761$

- Calculate consistency index (CI):

$$\text{Consistency Index (CI)} = \frac{(\lambda_{mean} - n)}{(n-1)};$$

where n is no. of sub – factors

$$CI = \frac{(4.187007761-4)}{3} = 0.06233592$$

- Calculate consistency ratio (CR):

$$\text{Consistency ratio (CR)} = \frac{CI}{\text{Random Consistency Index (RCI)}}$$

$$= \frac{0.06233592}{0.9}$$

$$= 0.069262134$$

Note:

- Consistency ratio (CR) for each pairwise should be less than 0.1.
- RCI value will vary according to the no. of sub-factors as appended in the Table below.

No. of sub-factors	3	4	5	6	7	8	9	10	11
Random Consistency Index (RCI)	0.58	0.9	1.12	1.24	1.32	1.41	1.45	1.49	1.51

This process is repeated for each factor and sub-factor and normalized weights are calculated and fed into the network for all the possible combinations of states assigned to the factors/sub-factors. The estimated weights are appended in

For example:

If EXP is high and training is good (AP is poor and no TA) then the weight or possibility of High Skill will become 0.774 and the same weight will be fed into the network.

Step 4: Analysis of different situations
After feeding the weights into the network, it can be used to analyze different situation. For example:

- One wants to see the effect of SKILL on RE. Give the evidence as SKILL is HIGH or LOW into the network and analyze the change in percentage of probability of high RE.
- Given the evidence as RE is HIGH or LOW, one can analyze the change in probabilities of each factor/sub-factor of being favorable or unfavorable respectively.
- Given the evidences according to the facilities available or not available in the organization, one can check positive or negative changes in the probability of high RE.

Similarly, the results are obtained for the examples as explained in the result section of this paper.

Pairwise Comparison Matrix for the Primary Factors

W.R.T. HUMAN ERRORS (HE)									
	SK	**TTGE**	**MN**	**EC**	**SL**	**SV**	**TX**	**nth root**	**WEIGHTS**
SK	1	3	3	5	5	5	3	3.191	0.349
TTGE	0.33	1	0.33	3	3	0.33	3	0.995	0.10
MN	0.33	3	1	5	5	3	3	2.164	0.236
EC	0.2	0.33	0.2	1	3	0.33	0.33	0.459	0.050
SL	0.2	0.33	0.2	0.33	1	0.33	0.33	0.335	0.036
SV	0.2	3	0.33	3	3	1	3	1.270	0.138
TX	0.33	0.33	0.33	3	3	0.33	1	0.726	0.079
Total								9.143	
λ MEAN 7.736222 **CI 0.122704** **CONSISTENCY RATIO (CR) 0.09**									

W.R.T. REPAIR COST (RC)							
	HC	SPC	SUC	AC	MC	nth root	WEIGHTS
HC	1	0.33	1	3	0.2	0.723324	0.111473
SPC	3	1	5	5	3	2.954177	0.455273
SUC	1	0.2	1	3	0.33	0.723324	0.111473
AC	0.33	0.2	0.33	1	0.33	0.372661	0.057431
MC	5	0.33	3	3	1	1.715321	0.264351
Total						6.488807	

λ MEAN 5.364636
CI 0.091159
CONSISTENCY RATIO (CR) 0.08

W.R.T. PREVENTIVE MAINTENANCE EFFECTIVENESS (PME)										
	TTGE	TX	MN	HE	SK	DC	TDAC	STD	nth root	WEIGHTS
TTGE	1	3	0.33	3	0.2	5	3	3	1.507	0.140
TX	0.33	1	0.33	3	0.33	3	3	3	1.142	0.106
MN	3	3	1	5	3	5	5	5	3.376	0.314
HE	0.33	0.33	0.2	1	0.33	3	3	3	0.814	0.075
SK	5	3	0.33	3	1	5	5	5	2.562	0.238
DC	0.2	0.33	0.2	0.33	0.2	1	0.33	0.33	0.314	0.029
TDAC	0.33	0.33	0.2	0.33	0.2	3	1	3	0.580	0.054
STD	0.33	0.33	0.2	0.33	0.2	3	0.33	1	0.440	0.041
T									10.73	

λ MEAN 8.929956
CI 0.132851
CONSISTENCY RATIO (CR) 0.09

W.R.T. MAINTAINABILITY (MN)						
	TX	**RSS**	**MD**	**AS**	**nth root**	**WEIGHTS**
TX	1	3	0.2	1	0.880112	0.169103
RSS	0.33	1	0.2	0.33	0.384162	0.073812
MD	5	5	1	3	2.942831	0.565429
AS	1	3	0.33	1	0.997491	0.191656
Total					5.204595	

λ MEAN 4.10819
CI 0.03606
CONSISTENCY RATIO (CR) 0.04007

W.R.T. EQUIPMENT EFFECTIVENESS (EE)						
	PME	**MN**	**BDF**	**UM**	**nth root**	**WEIGHTS**
PME	1	3	0.33	0.33	0.756027	0.149917
MN	0.33	1	0.2	0.33	0.384162	0.076178
BDF	3	5	1	3	2.59002	0.513589
UM	3	3	0.33	1	1.312771	0.260317
Total					5.042981	

λ MEAN 4.187008
CI 0.062336
CONSISTENCY RATIO (CR) 0.069262

W.R.T. SKILL (SK)						
	EX	**AP**	**TA**	**TN**	**nth root**	**WEIGHTS**
EX	1	5	3	3	2.59002	0.513589
AP	0.2	1	0.33	0.33	0.384162	0.076178
TA	0.33	3	1	0.33	0.756027	0.149917
TN	0.33	3	3	1	1.312771	0.260317
Total					5.042981	

λ MEAN 4.187008
CI 0.062336
CONSISTENCY RATIO (CR) 0.069262

W.R.T. TESTING TOOLS AND GROUND EQUIPMENTS (TTGE)							
	EC	**MF**	**PT**	**AR**	**STD**	**nth root**	**WEIGHTS**
EC	1	0.33	0.33	0.2	0.2	0.337146	0.051431
MF	3	1	1	0.33	0.2	0.723324	0.110343
PT	3	1	1	0.33	0.2	0.723324	0.110343
AR	5	3	3	1	3	2.667269	0.40689
STD	5	5	5	0.33	1	2.104189	0.320993
Total						6.555252	

λ MEAN 5.384657
CI 0.096164
CONSISTENCY RATIO (CR) 0.085861

W.R.T. ENVIRONMENTAL CONDITIONS (EC)					
	TAH	**ERG**	**PPE**	**nth root**	**WEIGHTS**
TAH	1	3	5	2.466212	0.63776
ERG	0.33	1	3	0.996655	0.257734
PPE	0.2	0.33	1	0.404124	0.104506
Total				3.866992	
λ MEAN 3.033029 **CI 0.016515** **CONSISTENCY RATIO (CR) 0.028473**					

W.R.T. REPAIR EFFECTIVENESS (RE)													
	EE	**SK**	**TTGE**	**SPQ**	**TX**	**PM**	**RC**	**MN**	**HE**	**EC**	**EX**	**nth root**	**WEIGHTS**
EE	1	0.33	3	0.33	3	3	3	0.33	3	5	3	1.5576	0.107682
SK	3	1	5	3	5	5	5	3	3	5	5	3.587195	0.247983
TTGE	0.33	0.2	1	0.33	3	3	3	0.2	3	3	3	1.110758	0.076787
SPQ	3	0.33	3	1	3	3	5	1	3	5	3	2.205779	0.152485
TX	0.33	0.2	0.33	0.33	1	0.33	0.33	0.33	0.33	3	0.33	0.426246	0.029466
PM	0.33	0.2	0.33	0.33	3	1	3	0.33	3	5	0.33	0.815205	0.056355
RC	0.33	0.2	0.33	0.2	3	0.33	1	0.2	0.33	3	1	0.52606	0.036367
MN	3	0.33	5	1	3	3	5	1	5	5	3	2.420461	0.167326
HE	0.33	0.33	0.33	0.33	3	0.33	3	0.2	1	3	0.33	0.636726	0.044017
EC	0.2	0.2	0.33	0.2	0.33	0.2	0.33	0.2	0.33	1	0.2	0.277752	0.019201
EX	0.33	0.2	0.33	0.33	3	3	1	0.33	3	5	1	0.90165	0.062331
Total												14.46551	

λ MEAN 12.48631
CI 0.148631
CONSISTENCY RATIO (CR) 0.098431

Index

Also of Interest

Repairable Systems Reliability Analysis
A Comprehensive Framework
By Rajiv Nandan Rai, Sanjay Kumar Chaturvedi and Nomesh Bolia
Published 2020. ISBN 978-1-119-52627-8

Most of the reliability literature is directed towards non-repairable systems, that is, systems that fail are discarded or replaced. This book is mainly dedicated to providing coverage to the reliability modeling and analysis of repairable systems that undergo failure-repair cycles. This unique book provides a comprehensive framework for the modeling and analysis of repairable systems considering both the non-parametric and parametric approaches to deal with their failure data. The book presents MCF MCF-based non-parametric approach with several illustrative examples and the generalized renewal process (GRP) based arithmetic reduction of age (ARA) models along with its applications to the systems failure data from the aviation industry. A complete chapter on an integrated framework for the procurement process is devoted to utilizing the concepts of multi-criteria decision-making (MCDM) techniques which will be of great assistance to the readers in enhancing the potential of their respective organizations. This book also presents FMEA methods tailored for GRP-based repairs. This text has primarily emerged from the industrial experience and research work of the authors. A number of illustrations have been included to make the subject lucid and vivid even for readers who are relatively new to this area. Besides, various examples have been provided to display the applicability of presented models and methodologies to assist the readers in applying the concepts presented in this book.

www.scrivenerpublishing.com

Printed in the USA/Agawam, MA
September 16, 2026

910661.055